DRIVEWAY DETAILING
Warrior

DIY Money-Saving Guide to Sports Car Detailing at Home on a Budget

S. L. Lucas

Copyright © 2022 by S. L. Lucas

All rights reserved. No part of this book may be reproduced or used in any manner without written permission of the copyright owner except for the use of quotations in a book review.

FIRST EDITION

Driveway-Detailing-Warrior.com

Book design by Publishing Push
Artist's name: Simon Heavens

ISBNs
Paperback: 978-1-80227-492-9
eBook: 978-1-80227-493-6
Hardback: 978-1-80227-494-3

Table of Contents

Introduction. v

Chapter 1
Sublime Safe Wash and the Love Connection 1

Chapter 2
Beadle's About - Indulgent Interior Detailing 15

Chapter 3
Stunning Stone Chip Repair 29

Chapter 4
Desirable DIY Detailing Products & Tools 39

    Part 1 - Detailing Hardware...39

    Part 2 - Detailing Liquids...51

Chapter 5
Wicked Wheels off Wheel-Arch Detail & Protection 63

Chapter 6
Wonderful Wheels off - Alloys, Brake Calipers & Exhaust Tip Detailing & Ceramic Coating. 83

Chapter 7
Exquisite Engine Detail and Protection. 107

Chapter 8
Perfect Paint Correction/Restoration & Protection: Methods for
Hand & Budget Machine Polishing. 123

Part 1 - The Paint Correction Decision 123

Part 2 - Paint Correction 'SOFT' 131

Part 3 - Paint Correction 'HARD' 140

Chapter 9
Handsome Headlight Semi-Permanent Restoration and Protection. . . 151

Part 1 - Headlight 'OFF' the Car Restoration 151

Part 2 - Headlight 'ON' the Car Restoration 164

Chapter 10
Ultimate Undercarriage Detail & Protection 175

Part 1 - Epic Under Trays Detailing 175

Part 2 - Uber Undercarriage Detailing 183

Acknowledgements. 197

About the Author. 199

Introduction

Perils of a Sloppy Ride

Did you know that a dirty, dingy car could be a *disaster* for your love life? I kid you not, a recent poll by a detailing products company revealed that over half (51%) of those polled would end a first date *early* if their date showed up in a dirty car. And 23% have even *finished* a relationship with someone because of their filthy, shameful vehicle! Flipping heck, this is tragic!

More than 1 in 10 Brits only wash their car if a passer-by writes 'Clean me' in the dirt on its body! I admit this happened to me years ago; I came out to find someone had written 'Please clean me' across my back window, and this *shamed* me into swift action. These days the usual comment passers-by make is more like '*Wow*! Please will you do mine next?' Sometimes I *do*; for a handsome reward, of course! That's to say nothing about a dirty car also being bad hygiene and potentially damaging to your health in these scary pandemic-ravaged days.

It's Not All About Looks

There is another important consideration, and many may think that keeping their car clean is purely aesthetic. Not so, my friends; *detailing* your ride can absolutely pay off financially in the long run. A detailing manufacturer's experiment added £2,000 ($2,694) to the appraised value of a BMW, using only off-the-shelf detailing products!

But What Is Detailing?

I've found that detailing can mean different things to different people. Some think it's a quick wash, vac and wax, while others believe it's squirrelling the car away in some high-tech lab for white-coated technicians to dismantle and cleanse every nut and bolt! Let's first agree that *Petrolheads* – car guys and girls – will usually be the ones who have the *passion* for making great home detailers.

The truth lies somewhere in the middle. Modern detailing is a combination of skills, passion, using the correct products, and going a step further to tackle areas you can't see, as well as those that you can. Perhaps you could describe detailing as the art of restoring the cosmetic appearance of a vehicle's interior and exterior, including paint correction and restoration, by hand, to like-new condition, just how she appeared in the sales brochure!

Good for the Soul

Okay, I'm going to ask you to keep an open mind here. I can hand on heart tell you that detailing your own cherished motor is not only good for your pocket, but it's also good for the mind, body and soul! Yes, plotting your detailing projects, being out in the fresh air, keeping busy and active, looking after your pride and joy, and enjoying the results is pretty darn rewarding for both the well-being and the wallet! Let's face it; there's plenty of time for vegetating on the sofa in front of the TV when it's dark outside or while your ride is safely stored away over winter!

As a confessed Petrolhead and Porsche enthusiast, I find no greater allure than cruising around in my gleaming, immaculately presented ride while observing (and secretly relishing and delighting in) the admiring, envious glances of pedestrians and fellow motorists alike.

Power of Home Detailing

Have you ever wished your handsome pride and joy, your Porsche or another sports car, still looked as dazzling as the day it graced the sales brochure? I certainly did, and I hope to take you on a journey of discovery to a place where you are empowered with the know-how to bring your beast back to her former glory with the power of home detailing.

The projects offered in this book provide a sure-fire way to keep the cost of ownership down while keeping your Porsche or other sports car fresh and good to go. I am by no means a motor mechanic, so I reckon if I can do a detailing project to an authentic high standard at home on my drive, anyone can do it! The satisfaction of completing a successful detailing project yourself using high-quality detailing products and essential home tools is blissfully rewarding.

DIY - Detailing It Yourself

There was something bugging me about the soaring cost of hiring professional detailers to bring out the full beauty in the sensual, flowing lines of my Porsche. Though my car was in excellent condition, I was quoted £1,231 ($1,650) by pro detailers for an exterior paint correction and ceramic sealant, £291 ($404) to have my alloys, brake calipers and exhaust tip ceramic coated, and £525 ($704) for undercarriage detailing and protection. Those three projects were quoted at a whopping £2,047 ($2,745) (the average of three separate quotes) from UK pro detailers in 2022 - *far* too much, methinks, and off-limits to the 'on a budget' sports car owner!

I drew on my experience of owning, detailing and driving three different Porsches since 2002: a 964 model 911, a 996 model 911 and my current 987.2 Cayman. I thought there *had* to be a foolproof way to do these projects at home and achieve the desired, *breath-taking* results, and there definitely is! I bring together in this book for your benefit, hopefully, in an engaging and inspiring way, all the pro hacks and detailing techniques I have learned along the way.

The Secrets Revealed

The eye-opening detailing projects explained in this book will take you fearlessly through the processes to achieve classy, professional standard detailing results for your Porsche (or any car for that matter). By following my simple, tried and tested step-by-step guides, you can keep every inch of your ride in *gorgeous* tip-top condition at a mere fraction of the cost charged by the pro detailers. And you will have the satisfaction of achieving those results yourself and knowing *you* completed your project correctly. Call us 'weekend warrior' or 'driveway hero'; the outcome is the same for the savvy home detailer.

So, with your valuable time, armed with the knowledge contained in this book. Let's get started transforming your Porsche or any sports car on a budget back to the *alluring* beauty she was in the sales brochures from new, turning her into the glorious envy of your friends and neighbours. With my help, YOU absolutely can *thrive* as a home detailer. So, let's do this Driveway Warrior style …

Chapter 1
Sublime Safe Wash and the Love Connection

In April 2011, I was free, single and bored at work one day. My thoughts drifted to this exotic Latin girl I had just texted, asking for a date - I know, the *model* of sophistication, aren't I? My phone chimed, it was *her*, and she accepted! *Brilliant!* Dinner at Las Iguanas, 8 pm in Brighton on Saturday. I'll pick her up in the Porsche and make a good impression. *Uh oh*, after that road trip to Beaulieu Motor Museum the previous weekend, the car was flipping *filthy* ... I must get that sorted before my date!

Bang for Buck Impact is what safe wash will give you. This chapter explains the safest, most effective way to wash your Porsche or other sports car to minimise the danger of introducing swirls and scratches to your *precious paintwork*. A relatively speedy path to a pretty darn sexy-looking motor at the start of your home detailing journey is well within reach; mark my words! At the same time, as your detailing skills and knowledge develop, you are beginning to unleash the potential of how your ride will look and feel to you.

Three days later, having spent the best part of a day detailing my GT Silver Porsche 996 model 911. I rang the doorbell to pick up my date. I'd even detailed myself - hair brushed, clean shaved, shirt ironed, jeans clean, new converse boots gleaming. If she was impressed by the car or me at the time, she kept it to herself but did admit much later she was *dazzled* - by the car!

Avoid the Roadside Car Wash

Unless you like having the same ancient, scratchy chamois leather used on your paintwork that's just been wiped over someone else's filthy wheels or having harsh traffic film removers sprayed all over your car, please avoid roadside or supermarket car park car wash operations. The same goes for those terrifying vast spinney, scratchy brush things at the automatic car wash. Detailers refer to these operations as *Scratch Wash and Swirl Factories*, and for good reason.

Sunshine - the Safe Wash Enemy

One of the most important things to remember if you want to avoid an enormous pain in the buttocks is NEVER to wash your car in direct sunlight. The sun will cause your cleaning products to dry on the paintwork before your rinse stage, leaving watermarks and streaks on the bodywork, which will have to be washed off again, and who needs that, right? The sun will also reduce the effectiveness of your cleaning products, forcing you to wastefully use more product, more water and electricity and, most importantly, more of your valuable energy and time. Granted, in the green and pleasant land of the UK, the sun is a novel pleasure we enjoy fleetingly from June to August. When you need to safe wash during these months, to avoid a torturous experience, I recommend you do so, if possible, before 9 am or after 6 pm, when your paintwork is in the shade and feels cool to the touch.

Splitting Up the Work

I find it easiest to split the tasks up, so I always do the interior separately. I cover this in the next chapter.

For Starters

> *Top Tip* First up, stick a beer or two in the fridge for when you finish, and grab a bottle of water or Thermocafé mug of coffee to keep you going.

Position your car, preferably with room to safely work around it. Remove your rings and watch, and make sure you aren't wearing anything zippy or with buttons or poppers that could scratch your car while you're working up close to it. Then get your Safe Wash Kit assembled, load up your Carry Caddy with your Spray Potions and Microfibre Cloths and keep it handy.

Maintenance Safe Wash Kit: Three Bucket Method

Detailing Hardware Required:

- Carry Caddy
- Nitrile Gloves
- Kneeling Mats
- Folding Stool
- Pressure Washer Kit & Extension Lead *or* Hosepipe and Nozzle
- Wash and Rinse Buckets with Grit Guards
- Wash Mitt
- Wheel Wash Bucket & Kit
- Car Blower
- Detail Guardz
- Handheld Foam Sprayer (diluted 1:10 Multi X to De-ionised water - if not using a Pressure Washer)
- Exterior Detailing Brush
- Microfibre Cloths (a bag full)
- Extra-Large Drying Towel
- Plush Buffing Towel
- Plastic Jug
- Window Blade/Squeegee

Top Tip Remember to rip off the tags from all your cloths and towels as these can scratch your paint!

Detailing Liquids Needed:

- Rust Repellent (RR) (diluted 1:3 RR to De-ionised Water)
- Bodywork Shampoo Conditioner
- Snow Foam (diluted 1:5 Snow Foam to Tap Water)
- Perl Dressing Solution Spray (diluted 1:3 Perl to De-ionised Water)

- All-Purpose Cleaner (APC) Spray (diluted 1:20 Multi X to De-ionised Water)
- Isopropyl Alcohol (IPA) Spray (diluted 70:30 IPA to De-ionised Water)
- Wax Spray
- Scratch Remover
- WD-40 Spray
- Microfibre Detergent

Let's Do This!

Step 1: Lift the wiper arms and open the petrol flap. Fill your Rinse Bucket with clean cold water. Seize your Wash Bucket add two capfuls of Bodywork Shampoo, and fill with warm water, swishing it around to mix well. Grab your Wheel Bucket, add a further two capfuls of Bodywork Shampoo and fill with warm water. If your wheels are really filthy, add 2 capfuls of Multi X APC for extra cleaning power.

Ensure all your buckets have their grit guards at the bottom. Prepare your Snow Foam Dispenser with a 1:5 mix (Snow Foam to Tap Water).

> *Top Tip* Double-check that the nozzle or foam dispenser has clicked correctly into position when attaching to your pressure washer lance. I've seen someone firing the nozzle into the side of their car, causing a sizeable crater (I promise it wasn't me)! Best avoid doing similar on your pride and joy.

Rustproof the Discs

Step 2: Don your Gloves, give each brake disc (rotor) a good spray down with Rust Repellent (RR) and allow to cure for a couple of minutes. This will help prevent your brakes from turning orange when the water hits them. Then, move the RR away from the car out of the way. Put your Detail Guardz in position under the outer edge of your tyres; you'll be thankful for them when using the hose and car blower. Check that your windows, boot and rear lid are closed tight, and manually raise your rear spoiler if applicable.

> *Top Tip* Roll the car a few inches to access the bit of the disc hiding behind the brake caliper.

Wheels, Tyres, Brake Calipers & Exhaust Tip(s)

Step 3: <u>Always</u> tackle your wheels *first;* they are the dirtiest part of the car and will sling muck onto your clean paintwork if you wash them after you've done the paintwork. If using a Pressure Washer, generously coat each wheel, brake caliper under the wheel arches and exhaust tip(s) with snow foam and leave it for two minutes. If not using a Pressure Washer, load up your Handheld Foam Sprayer with a 1:5 mix (APC to De-ionised Water), shake well, and liberally foam up each wheel, brake caliper and exhaust tip(s).

Then take your Wheel Wash Bucket & Brush Kit and, using your Kneeling Mats and Stool if you wish, get scrubbing! Start with the Meguiar's Supreme Brush, regularly rinsing it in your Wheel Wash Bucket, and work your way through your Wheel Brush Kit, keeping them super-well-lubricated with wash water (a dry brush is a scratchy one). Reach through the spokes and work the bristles over the calipers as best you can. Use the Wheel Bucket Wash Mitt to clean around the back of the spokes. Reach under the wheel arch lip roll to clean out the debris that collects there, or you'll always know grime is lurking there!

Take the opportunity to inspect your rims for any dreaded 'kerb rash.'

Finish by attacking the tyre walls with your Tyre Brush, rinsing the brush regularly in your wheel bucket, keeping it well lubricated, and your tyres will soon be spotless. Be careful not to use the tyre brush on your alloys or anywhere else, as it will scratch.

Now is an excellent time to tackle your exhaust tip(s) with your Wheel Brush Kit to get them sparkling. Take the opportunity to examine your tyres for any concerning splits or damage that may need attention. Next, give the wheels, brake calipers, under the arches and exhaust tip(s) a good rinse with your pressure washer or hose nozzle. Ensure your Pressure Washer attachment is the wide fan nozzle and keep the powerful spray <u>moving</u>. You are done with your wheels and exhaust tips for now: no need to dry them at this stage.

Just move your Wheel Wash Bucket & Brush Kit away from the car out of your way.

Wipers

Step 4: Take your IPA Spray and a clean Microfibre Cloth. Soak an edge of the cloth with IPA, pinch the wet Cloth around your window wiper rubbers and give them a thorough clean, moving to a new damp part of the Cloth as you go.

This is the absolute best way to keep your wiper rubbers squeaky clean and performing to their optimum.

Bodywork Pre-Rinse

Step 5: Now, take your Pressure Washer with the wide fan nozzle or hose nozzle and rinse the car down thoroughly with water. Focus more of the spray on the lower third, where road grime tends to be heavier. Keep the spray moving over your paintwork as a precaution against causing damage.

Bodywork Snow Foam Pre-Wash

Step 6: Then re-attach your Snow Foam Lance to your Pressure Washer and spray the whole car with thick, satisfying soapy suds, ensuring complete coverage top to bottom (except the wheels). If not using a Pressure Washer, load up your Handheld Foam Sprayer with a 1:5 mix (APC to De-ionised Water), shake well, and lay down a foam covering over the whole car concentrating the spray more on the lower third. You'll probably need a couple of loads with the Handheld to get a decent foam covering.

Step 7: With the Snow Foam or APC Foam coating the car, now is the time to go in with your Large Sash Detailing Brush. Fill your Plastic Jug with clean soapy wash water, and work your way methodically around the car. Work the brush to agitate grime away from panel gaps, door handles, badges, emblems, grills, lights, body trim, wiper arm springs and inside the petrol flap. Keep the brush well rinsed and lubricated with wash water as you go. The idea behind the jug is that you don't have to keep returning to your wash bucket to rinse and lubricate your brush!

Don't be tempted to use your Wash Mitt at this stage, as you will succeed only in grinding dirt suspended in the Snow Foam or APC Foam into your paintwork! The Wash Mitt is deployed at Step 8.

Top Tip I like to place four mats, one front, one rear and one on each side of the car as I move around detailing down low, for which my knees always thank me!

Next, take your Pressure Washer or Hose Nozzle and rinse the car down thoroughly to remove all foam residue. Finish this step by discarding the contaminated jug of water, don't tip it back into your wash bucket!

Contact Wash

Step 8: This is your primary bodywork contact wash stage. With clean wash and rinse water buckets to hand, dunk your Wash Mitt in the clean wash water. Work methodically from top to bottom, gliding the Mitt in straight lines, back and forth using light pressure, avoiding rubbing or circular motions. Make sure you regularly rinse your Mitt thoroughly in your rinse bucket, squeezing it out into the rinse bucket before dunking the clean Mitt back in your wash bucket. This is how to help avoid spreading road contamination back on the car and grinding it into your paint. As you get to the lower third of the vehicle, increase the number of times you rinse your Mitt. When finished with the Mitt, move the Wash and Rinse Buckets away from the car out of the way.

Step 9: Once again, take your Pressure Washer or Hose Nozzle and thoroughly rinse the car down, from top to bottom, to remove all the soapy wash water.

Contactless Drying

Step 10: I've heard it said that up to 80% of swirl marks are introduced to paintwork at the drying stage – a frightening prospect, right? While this may be the case at the local scratch wash operation, fear not, for Driveway Warriors know better! I always start the contactless drying stage by opening and firmly closing the car doors several times to expel some trapped rinse water from the door recesses. Granted, in doing so, I've had the odd, strange look from passing pedestrians, but hey, it works for me! After that, grab your Car Blower, and trail the extension lead over your shoulder so it doesn't drag scratchily over your paintwork. Go around the car, air blasting rinse water out of all the nooks, crannies and crevices, including your wheels and exhaust tip(s).

Contact Drying

Step 10A: The contact drying stage (if you _don't_ want to apply spray wax).

Grab your Squeegee and go around the car, sliding the water down off the glass. Don't use the Squeegee on your paintwork as it can cause scratches. Then, take your large, dedicated drying towel, spread it flat on the wet roof, and draw it across the top, back and forth, without applying pressure. This should be enough to dampen the towel and 'prime' it. The reason for doing this is that the towel works best when slightly damp.

Next, go over the whole car with the towel, including the glass, applying only light pressure and folding the towel as you go; definitely don't rub, as this may cause swirl marks in your paint. The towel is epically absorbent, so you shouldn't need to wring it out at any stage. Now is the time to keep a keen eye out for any scratches, scuffs, or stone chips lurking on the paint that may need scratch remover or paint-stick attention later.

Open all your doors, bonnet lid and boot lid. Take a clean microfibre cloth and go around the car, drying any remaining rinse water from the door shuts and the upper and lower areas of the boot and bonnets lids.

Top Tip Assuming your door windows are of a 'framed' design, i.e., not frameless like my Cayman, wind the windows down about 8cm (3in) to dry the top part of the door glass to give a seamless professional finish.

Waxing

Step 11: The contact drying stage (if you *do* want to apply spray wax).

Go on; you've come this far. It seems rude not to treat your ride to some heavenly wax-loving! Take your Spray Wax, shake well, and work methodically, one panel at a time, from top to bottom. Do 2-3 sprays per panel onto the still wet paintwork, spread with a clean Microfibre Cloth, and buff to a glossy mirror finish with your separate dedicated, plush Glart Buffing Towel.

This Spray Wax really is a doddle to use because it performs like a drying aid, giving waxy lubricity under your Buffing Towel, enabling it to glide across the whole car with ease. You can use it anywhere except on your glass. Less wax is more, don't spray too much on the car, or it can smear and become a chore to buff off. However, if you do get a little too much wax on the paint, just spritz the overloaded area with IPA, wipe it down with a clean Microfibre Cloth and carry on. Use a separate clean Microfibre Cloth to dry the glass.

Wax up Your Wheels

Step 12: Time to buff up your wheels and exhaust tip(s) now.

Swap your Glart Buffing Towel for a couple of different coloured clean, dedicated, wheel Microfibre Cloths. Spritz your alloys and brake calipers with spray Wax, avoiding getting too much on the brake discs (rotors). Then spread the Wax with one microfibre cloth and buff to a shine with the other.

Do the same on your exhaust tip(s) for that fresh, professionally detailed look. I cover ceramic coating the wheels, calipers & tips in a later chapter.

Dressing & UV Protection

Step 13: Now to get some ever so indulgent dressing on those plastics and rubbers! I tend to keep two bottles of CarPro PERL dressing prepared and labelled: one diluted 1:3 (Perl to De-ionised Water) for exterior plastics and rubbers, the other diluted 1:2 for tyre walls and engine bay. On my Cayman, I dress several areas:

- The front lower spoiler lip
- The scuttle panel at the windscreen base
- The tyre mud flaps
- The door rubbers
- The rubber weathering strips embedded in either side of the roof
- Last but not least, the oft-forgotten inside the petrol flap

To prep for dressing, take your IPA Spray diluted 70:30 (IPA to De-ionised Water), spray down the part to be dressed and wipe down thoroughly with a clean Microfibre Cloth. Don't worry about getting IPA Spray on your paint; it will quickly flash away to nothing with the help of your drying cloth. Next, take your CarPro Perl spray, diluted 1:3, spritz over the part to be dressed, and then work it in with a clean Microfibre Cloth. Depending on the level of darkening and sheen you prefer, you may want two applications of Perl, waiting ten minutes for the first coat to cure before applying a second coat. The Perl also adds a welcome UV protectant to any surface it's administered to.

Moving on to the tyre dressing, take your CarPro Perl spray diluted 1:2 and evenly spray a coating right around the tyre wall. Then, using an old Microfibre Cloth, gently work the dressing into the tyre wall. It will give you a lovely, durable mid satin sheen that will last several weeks, and I've found it will not sling all over your paintwork as soon as you drive the car.

More Detailing Hacks

Step 14: We are 'detailing' right, so I have a couple of cool little hacks to finish with. Here's a sad anecdote about a fellow Porschephile who splashed out the princely sum of £75 ($101) on four sexy Porsche OEM metal dust caps, only to have them rapidly seize on the tyre valves. The unfortunate fellow was gutted, to say the least, when his local indie (Independent Porsche Specialist) had to butcher the pricey items to get them off!

You can avoid this happening to you by removing your metal dust caps after every safe wash. Blow out any water, squirt a drop or two of WD-40 inside the dust cap and replace them on the tyre valves.

Next neat hack. Keep your wipers tight to the glass and working at an optimum level by wrapping an old microfibre cloth around the wiper base to protect the car from overspray. Douse the wiper arm spring with WD-40 and work the wiper arm up and down a dozen or so times to help the WD-40 penetrate.

These couple of actions take mere minutes after every safe wash and will serve you well for months of motoring; it's all in the 'detailing,' right!

Photo Time

You are DONE, stand back and admire your wondrous creation!

At this stage, you might fancy taking a few snaps for posterity (or to keep for when you want to move the car on). Many of the blue Cayman pics in this book were taken on my inexpensive smartphone.

You don't need to be David Bailey to take decent images, though you do need to think about the right time of day to do it. Your freshly detailed beast is basically a vast mirror that will mercilessly reflect sunlight, which will destroy the quality of your shots. The best time for your photos is just after sunset or before sunrise. Wherever you are, Mr Google will inform you what time of day these are due.

Try and position the car where you have enough room to move around it to get a good angle for your snaps. Clear all your detailing gear, the dustbin and the recycling bin away from sight. Think about your camera height; by that, I mean don't just take all your snaps at standing eye level. I think some of the coolest, most purposeful shots are taken low down to the ground, giving your ride a menacing stance! I did use a smartphone tripod for my lowdown shots, which cost me just £5 ($6.62) from eBay. Give it a go, and you might be impressed with what you can achieve!

Trust me; these pictures will sustain you if you store your ride over winter. As well as help you get top dollar when you decide to sell the car.

Clear Up

Now for the part everyone hates! Clear up is a necessary evil and will ensure all your precious detailing kit stays in tip-top condition, ready to go next time you need it.

Start by returning all your liquids and spray bottles to storage, but do check all lids and spray heads are secure and keep an eye on any dilutables that need topping up. If you used the Handheld Foam Sprayer, rinse it out with clean water, semi-pressurise it and spray out some clean water to clear the feed pipe and nozzle. Give it a quick dry and store it away. Grab your Car Blower, Caddy, Mats, Stool, Detail Guardz and Squeegee and give them a quick wipe down if necessary to dry and store them away.

Discard all the bucket water (my rinse water goes in the water butt; the garden loves it). Give your three Buckets and Grit Guards a rinsing blast with the Pressure

Washer or Hose Nozzle, then put them somewhere to drain upside down. Hold back the wheel bucket as you'll need it in a bit.

Next, gather all your used Cloths, Mitts and Towels and squeeze them out if necessary. If the exterior Cloths are very mucky, chuck them in a clean bucket half-filled with clean, warm water, add a small squirt of washing-up liquid (dish soap) and give them a short pre-wash with your gloved hand. Then squeeze them out and pop them in the washing machine at 40C (104F), add the Microfibre Detergent, and stick on a 30-minute cycle with a gentle spin.

The only thing I wash separately from the other cloths is my Glart Soft Buffing Towel; it's a bit more delicate, so I do it at 30C (86F). When you take them out of the machine, give them a good shake to re-set the fibres and hang everything on a clothes dryer to dry naturally, using pegs for the mitts.

Top Tip I keep my Microfibre Detergent with my detailing kit; otherwise, it gets used for the household laundry!

You should now be left with just your Pressure Washer, Brushes and Wheel Bucket. Dump all your Brushes in the Wheel Bucket and give them a quick rinsing blast with the Pressure Washer or Hose Nozzle. Discard that rinse water, add a squirt of washing-up liquid to the Bucket and half-fill it with hot tap water, swirling it around to mix. Washing-up liquid is a superb degreaser and is perfect for brush cleaning in this way. Spend a few minutes with your gloved hands in the soapy water, massaging the brushes clean. You'll find the Tyre Brush gets the muckiest, so I give this one a small extra squirt of washing-up liquid and work it in using the hot water. Then dump the soapy water and give the Brushes a final rinsing blast in the Bucket with the Pressure Washer or Hose Nozzle and discard the rinse water.

I give all the Brushes a good wrist flick to fling off some of the excess rinse water. I then lay them on an old dry towel under the radiator to dry or in the sun if you're lucky. Give your three clean Buckets and Grit Guards a quick wipe to dry and store them away with their lids on to keep the insides clean. Give your Pressure Washer and accessories, including the power cable, a wipe down to dry and store them away. The next day when your Towels, Cloths, Mitts and Brushes are fully dry, store them away, keeping your Wheel Brushes in your Wheel Wash Bucket.

Reflection

At last, you are done. Congratulations, phew! Not only does your ride look magnificent, but you've also looked after your kit. You can smile knowingly at having avoided the local roadside scratch-wash operation and having saved £180 ($243) on the cost of a professional detailer doing the same job you've just done. Smile even more broadly in the knowledge that your saving on detailing costs will spiral over time with each detailing project you and your kit undertake.

Put your feet up and have a good rest, crack open that icy beer or whatever is your pleasure and plot your next Driveway Warrior, money-saving detailing project with my help ...

Let's go back to that first date I mentioned at the start of the chapter. Of course, I'm not suggesting she married me a year later because of my exquisitely detailed Porsche, at least I hope not, but it certainly didn't hurt and was a cracking start to the date! It set the tone for what was to come, and I choose to give at least *some* credit to the *power* of home detailing and my sexy motor!

Chapter 2
Beadle's About - Indulgent Interior Detailing

Christmas Day 2021, approaching midnight, this sober, on taxi duty Driveway Warrior dutifully sits in his VW Golf, warming up the heater, waiting for the overfed, slightly inebriated Mother and Wife to exit my sister's house, where we'd been celebrating the festivities. As the freezing rain lashes down on the windscreen, I'm tired but happy and looking forward to my warm bed.

Mrs *Senior* and *Junior* totter unsteadily down the driveway towards me, clutching onto each other, shoulders hunched against the driving rain. It's a *stinker* of a night. In a minute, they'll both be safely in the car, and we'll head home. I'll drop Mum off, no drama whatsoever - or *so I thought!* However, what unfolded next was a slapstick scene from a *Carry On* film.

Instead of walking *around* the rain-sodden, mud-caked grass verge between house and car, Mrs Senior and Junior *inexplicably* stepped directly onto the perilously saturated bog. A second later, Mrs Senior landed flat on her back in the squelching mud, followed instantly by Mrs Junior going down on *her* back right beside her mum-in-law. There they lay, flailing around in the dirt, squealing in misery, literally bathing in the slime.

Senior writhing around, glazing one festive M&S garb and Junior, generously coating a carefully-selected designer outfit in thick, filthy *viscus* muck - oh dear, sweet Jesus.

"Oh my God!" I shouted in *horror*; this can't be happening.

In a second, Jeremy Beadle will pop up from behind a bush, microphone clutched in withered hand, thrusting it toward my face. I threw the car door open and stumbled out, hauling the two bedraggled, filth-encrusted ladies to their feet, becoming mud blanketed myself in the process.

"Is anyone hurt? Let's get you both home," I croaked sympathetically.

Thankfully, the only hurt things were pride and designer labels. Oh, and my immaculate interior: the leather, carpet mats, dash, upholstery, seatbelts, steering wheel, switches, gear knob, even the *flipping* headlining, all covered in a layer of stinking mud. Oh Lord, it's enough to make a man *weep!*

Boxing Day

As we unhappily begin the clean-up, I'm generally rationalising to Mrs Lucas that your interior, cockpit, or cabin, call it what you will, is the place where you spend most of your time while enjoying your car. So, it follows that your interior should feel as inviting, fresh, and pleasant as possible, right?

"Yes, dear," she replies dejectedly.

I know when to *zip* it!

A freshly detailed cockpit always seems to uplift me when I'm whizzing around enjoying my ride. In my humble opinion, the smells and textures of the leather, the gleaming trim and the sparkling windows are all rather glorious to the senses. Sure, I'm waxing lyrical here, but I reckon any *Petrolhead* will understand where I'm coming from.

Breaking up the Tasks

I like to separate my interior detailing and safe wash sessions and do them on different days. That way, I can focus on each element without rushing or feeling overwhelmed by the task list! We're home detailers, so we don't really need to think of detailing as a business in terms of how many maintenance washes we can plough through in a day. Who knows? Maybe the business angle of home detailing could be the next book!

Timing the Weather

With interior detailing, you have more flexibility with weather conditions than a safe wash. I still recommend picking a dry, mild day when your interior finishes are cool to the touch. Detailing products don't usually react well to being used on any hot surfaces.

Degrees of Dirt

While your exterior safe wash process remains consistent, interior detailing differs. Your process and amount of time and work required are determined by the grime level in your interior. You'll need to assess if your carpets just need a good vacuum, or do they need a complete shampoo treatment? Does your trim need a standard spruce up, or is it crying out for a deep dive clean? Your eyes and, to some extent, your nose will give you the answer! This section assumes that your interior is typical to moderate muck level.

Interior Detailing Kit List

Detailing Hardware Required:

- Carry Caddy
- Nitrile Gloves
- Kneeling Mats
- Folding Stool
- Vacuum Cleaner & Extension Lead
- Microfibre Cloths (a bag full)
- 'Dash' Detailing Brush
- Magic Eraser
- Upholstery & Leather Care Brushes
- Wheel Brush (for pedals)
- Hanging Air Freshener
- Clip Hangers x 4
- Handheld Foam Sprayer (1:20 APC to Water) for the Mats

Detailing Liquids Needed:

- All-Purpose Cleaner (APC) Spray bottle - diluted 1:20 Multi X to De-ionised Water
- Perl Dressing Spray (diluted 1:5 Perl to De-ionised Water)

- Glass Cleaner
- Fabric Protector
- Handheld Foam Sprayer (diluted 1:5 Multi X to De-ionised Water)
- APC Spray (diluted 1:5 Multi X to De-ionised Water for the pedals)
- WD-40 Spray
- Perl Dressing Spray (diluted 1:3 Perl to De-ionised Water for the rubbers)
- Microfibre Detergent

Preparation

First up, may I suggest putting some beers on ice, ready for when you finish and grab your water bottle to keep you hydrated while you work.

Position your car, preferably with room to safely work around it. Make sure all your Brushes are clean, then get your interior detailing kit assembled. Load your Carry Caddy with your Spray Bottle potions, Brushes, and Microfibre Cloths, and have your Caddy handy. Lose all your rings and watch, and make sure you aren't wearing anything zippy or with buttons or poppers that could scratch your ride while you're working in and around it.

> ***Top Tip*** Use your colour-coded Detailing Spray Bottles to help distinguish between your interior and exterior sprays. I keep my interior APC Spray (diluted and labelled 1:20) in a red spray head bottle and my exterior APC Spray (diluted and labelled 1:5) in a blue spray head bottle. Just about foolproof for me because, despite the labelling, I associate the 'red' with my interior red stitching trim and the 'blue' with my exterior Aqua Blue paint finish - brilliant!

Get it all Out

Step 1: Open all your doors, front boot (frunk), rear cargo lid, and remove everything from the car. Go through your door pockets, cubby holes, glove box, and deposit everything in a bag, away from the car somewhere safe. If you have faith in your battery, stick on your favourite radio station while you work.

Floor Mats

Step 2: Gloves on, remove your floor mats from the car, shake them out and give them a few whacks against the side of the house to loosen any ingrained dirt. Assess their condition but don't put them on the ground as they'll just pick up dirt.

Top Tip Interlock a couple of your brilliantly useful kneeling mats, dump your car mats on top, and give them a good vacuum front and back using your crevice tool. If they're filthy, seize your Handheld Foam Sprayer, shake well and cover the mats with the satisfying soapy foam. Allow to penetrate for 30 seconds, then go to work with your upholstery brush, working back and forth, vertically and horizontally, covering every inch.

Mat Pattern Striping: If you fancy giving your mats a professional detailing style finish, why not pattern stripe them? It's super easy if you know how and looks rather cool! While the mats are still damp from the foam treatment, take your Upholstery Brush and, starting from one end, carefully brush all the fabric in one direction across the mat, so the fibres lay flat. Now you can create your stripes by holding your Brush straight and carefully brushing the fibres in the opposite direction, so they stand up. Work your Brush from top to bottom, then on the next line, work your Brush from bottom to top in the opposite direction. Repeat across the whole mat alternating which way you brush the stripe.

Once you're happy with your stripes and your mats are clean and smelling fresh, finish by taking your Fabric Protector and evenly spraying on a thin coat in a crosshatch pattern. I usually leave the Protector to cure for a few hours before using the mats.

Top Tip A *smart* way to dry your floor mats is by hanging them somewhere convenient from a 'clip cloths hanger' (a clothes hanger with two sturdy clips at the bottom).

Vacuuming

Step 3: I like to think of my knees at this point, and I place a kneeling mat on the floor at the appropriate area on either side of the car. Sometimes I even park my posterior on my stool to give my knees a break! Grab your Vacuum Cleaner and start in your cabin with a bit of multi-tasking (yes, blokes can multi-task)! Take your 'Dash Detailing Brush' in one hand, your Vacuum Crevice Tool in the other, and holding the Crevice Tool underneath your Brush, start to agitate dust out of all those tricky to reach dash and trim areas around switches, screens, vents, seams, buttons and knobs!

Move the seats back as far as they'll go so you can vacuum underneath them, then as far forward as they'll go, so you can reach under the backs. Next, utilising your

Crevice Tools, go over the lower areas in the nooks, crannies and carpeted areas. Even the door bins, glove box, gear stick gaiter, pedals and the space beneath them should be subjected to your vacuuming *frenzy!*

Top Tip If dirt and debris on your carpet resist the assault of your vacuum crevice tool, seize your clean ValetPro Wheel Brush and agitate the stubborn debris with the bristles. Any ingrained debris should then more easily succumb to your vacuum.

Seats: Don't forget the seats where you sometimes need to ease the seams apart to run your Dash Brush along the gaps, followed by your Vacuum Crevice Tool.

Top Tip Reserve your Gyeon Leather Care Brush for any stubborn stains on your leather seats. Spritz some APC Spray (1:20 mix) onto the Brush, gently work the bristles over the stain, blot with a clean, Microfibre Cloth and check the results.

If the stain persists, spray a small amount of APC directly onto it and another spritz on your brush and go again, working the bristles gently over the mark. At the very least, you can safely reduce the blemish in this way, then blot the area with a clean Microfibre Cloth.

Front Boot (Frunk) & Rear Cargo Space: Turn your attention to the front boot (frunk) and the rear cargo area when finished in the cabin. Be gentle with how you move your Crevice Tool across the delicate rubber type finish found in some Porsche models (including the Slate Grey interior in my Cayman). If you're working on a hatchback model, when vacuuming the boot, you may find it easier to push the back seats forward and down to improve access for vacuuming. When you have finished, move your Vacuum Gear and Dash Brush away from the car out of your way.

Glistening Glass

Step 4: I always do the glass at this stage because if any overspray settles on the interior surfaces, it can be removed at step 5. Shake your Glass Cleaner well and spritz the inside of the glass, taking care not to get too much on the trim. Take a clean, folded Microfibre Cloth and clean right around the border of the glass. Then fold your Cloth and attack the glass vertically and horizontally, ensuring you cover every bit. The windscreen can be awkward to reach, and I find that the folded Cloth on the back of the hand helps here. Don't forget the rearview mirror and sun visor mirror glass. When finished, move your Glass Cleaner and used Glass Cloth away from the car out of the way. I tend not to replace them in my Caddy at this stage, so I know everything remaining in the Caddy is yet to be used.

Roof/Head Lining

Step 5: Assess the condition of your roof lining. If it's a bit grubby, you can spot clean it. However, you need to be careful because the roof liner is rather delicate, it doesn't like to get too wet, and it can be prone to sagging if not treated with respect. If the roof liner isn't a super delicate finish, such as Alcantara, you can have a go at improving it. Take your clean Dash Detailing Brush, spritz it with APC (1:20 mix) and gently work the soft bristles over the grubby area. Blot with a clean Microfibre Cloth and check your results. If the stain persists, you can go again with a second gentle pass of brush and APC and finish with a final blot with a clean, dry Microfibre Cloth.

All-Purpose Cleaner

Step 6: Next up, starting in the cockpit, take your APC Spray (1:20 mix), shake well, and grab a handful of fresh Microfibre Cloths. Fold a Cloth, moisten it with APC and methodically wipe down all your surfaces top to bottom, folding and replenishing the Cloth with APC as you go. Don't forget the insides of the doors and the often-overlooked visible bit of the seatbelts and backs of the sun visors. Unless your steering wheel is a delicate Alcantara finish, you can give it the same APC and cloth treatment. By spraying onto your Cloth and not directly onto the surfaces and trims, you avoid overspray and help to preserve your delicate instruments, electronics and trim.

Unsightly Scuffs: By this stage, you've had a chance to inspect the condition of your interior. Are there any pesky scuffs or kick marks letting the side down? These marks accumulate on the insides of the doors (door cards) and on the plastics on the bottom of the door aperture (the bit you step over to enter and exit). Be mindful of delicate finishes, and you can improve these bothersome bits. Take your Magic Eraser, moisten a corner with APC (1:5 mix), and work it using medium pressure over the interior scuff in a crosshatch pattern, keeping the Eraser lubricated with APC. You should see a marked improvement as the Eraser lifts the scuff material away from the surface. Wipe it down with a clean Microfibre Cloth and finish off your fix with a squirt of dressing, worked in with another clean Microfibre Cloth. This final fix works particularly well on those textured interior panels and surfaces where a Microfibre Cloth may be less effective.

Back to the Front Boot and Rear Cargo: Changing Cloths if necessary, move onto the front boot (frunk) and rear cargo area and give those finishes the same treatment with the APC Spray and Microfibre Cloth. Check if any scuff marks lurk in areas that may benefit from an attack with the Magic Eraser treatment. Next, move the APC Spray, used Cloths and Eraser away from the car out of the way, ready for the next step.

Dressing

Step 7: Now that your interior is squeaky clean, it's time to lay down some lovely mid-satin sheen dressing. Take your Spray Bottle of diluted yet heroic CarPro Perl, shake well, and clutch another batch of clean Microfibre Cloths. Fold a Cloth, moisten it with a few squirts of Perl and apply the dressing to all your surfaces, working methodically top to bottom, folding and replenishing your Cloth with Perl as you go. I think you'll find this to be the most satisfying stage of your interior detail. You'll see the magical but subtle, darkening, restorative effect almost immediately, and it even adds UV protective qualities in the process! I've always found that just one coat of this treatment is enough to produce the desired finish, i.e., not overly glossy, and it will last up to 3 months before needing a top-up.

Next, take a fresh Microfibre Cloth, moisten with Perl and give your leather seats the same treatment, working the Perl into the creases and folds of the leather as you go. The Perl will nourish the leather, giving your seats the same UV protection as your plastics and trim. Apply a little more of the dressing into areas where the leather is more 'aged,' then stand back and admire how fresh and strikingly younger your interior looks.

However, you are not quite finished.

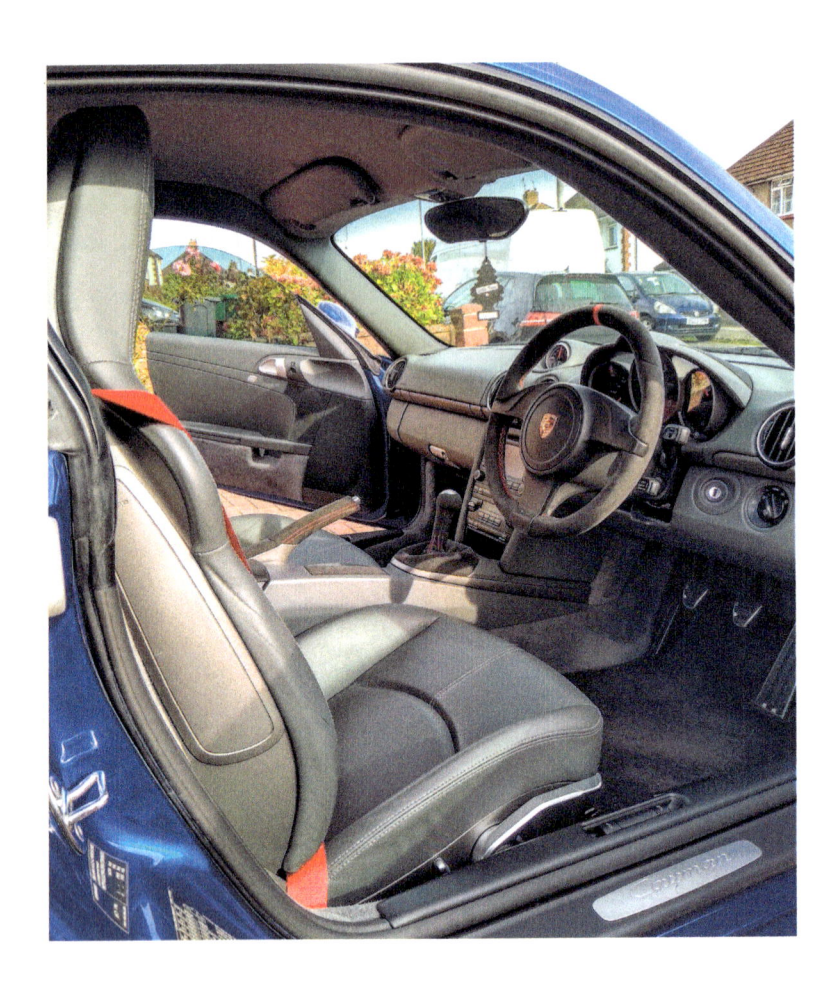

Finishing Touches

Step 8 - Pedals: The last detailing task for your interior may surprise you. It will really help lift the overall feel to a professional standard, as well as keep your pedals nice and tactile underfoot during your rapid B-road adventures! You need to detail your pedals – who knew that was even a thing? So, they should already have had a good vacuum; the next thing is to lay down several old Microfibre Cloths on the carpet underneath to protect this area. Then load your caddy with your APC Spray (diluted 1:5 Multi X to De-ionised Water), a couple of clean Microfibre Cloths, and your clean ValetPro Wheel Brush.

With your Gloves on, hit your pedals and footrest with APC Spray and allow to soak for 30 seconds. Then moisten your Wheel Brush with more APC and work it methodically all over your pedals and footrest one by one. Work the bristles into the textured grooves, the stem and the backs of the pedals. Then wipe everything down with a Microfibre Cloth, folding it as you go. Give the Brush a quick clean with the Cloth as well, ready to repeat the whole process until you've done two complete passes with the APC and the brush and wiped everything down again with the Microfibre Cloth.

Finish by giving the pedals and footrest a third APC Spray down and this time, take a clean Microfibre Cloth and work it over every inch, buffing and folding the Cloth as you go. You have DONE a brilliant job - who would have thought a fresh-looking pedal set could bring such contentment? Word of warning: DO NOT be tempted to apply any type of dressing to the pedals as dressing contains silicone, which will make your pedals slippery. This would be a terrible idea, not to mention highly dangerous.

Hinges and Rubbers: No, you haven't strayed into a chapter of *'50 Shades of Grey'* (not that I've read it!), though I promise you have almost finished. Now is an excellent time to look after all your hinges and associated rubber hoses, which is much easier than you might think.

Retrieve your APC Spray (1:5), Perl Dressing Spray (1:3), a couple of clean Microfibre Cloths, and your clean ValetPro Wheel Brush.

Open all your doors, front boot (frunk), rear cargo lid and petrol flap. Start at the side doors and spritz some APC over the door hinges and any rubber hoses lurking there. Spritz more APC on your Wheel Brush and go in with it, working the bristles all over the hinges and rubber hoses. Next, take a clean Microfibre Cloth and wipe down the hinges and the rubber.

When it looks clean, grab your WD-40 Spray, wrap an old Microfibre Cloth around the hinge to protect it from overspray, and spritz down the hinge enough to cover it. Then push the door open and closed half a dozen times to help the lubricant penetrate, remembering to smile politely as passers-by gaze at you with curiosity! Finish by hitting the rubber hose with a squirt of Perl dressing (1:3) and work the dressing into the rubber with a clean Microfibre Cloth. The Perl will restore and nourish the rubber, giving a reliable mid-satin sheen and prolonging its life while providing UV protection. Repeat this process where necessary on your front boot, rear cargo lid and petrol flap.

Air Freshener: Surely, it must be illegal to complete an interior detail without hanging a new air freshener from your mirror! My favourite is *'Little Trees Black*

Ice,' a black Christmas tree-looking shape with a rather masculine aroma of woods and citrus! The best way to use these is not to rip off the enclosing bag, just snip open the top of the bag and pull the tree out about 1cm (0.5in), then pull the tree out a little more week by week. This way, it'll last much longer and give you a subtle pleasurable hit rather than an assault on the nostrils!

Reset: Finally, replace your dry and protected floor mats back in the car and return the seats, steering wheel, rearview mirror etc., to their correct position.

Photos

If you've already read Chapter 1, you'll know all about the benefits of capturing snaps of your freshly detailed pride and joy, looking its most magnificent! This also applies to your interior; the principles applied to your exterior shots are the same for your interior. So, get all your doors, front boot and rear cargo lids open, fully recline your seats and push them back as far back as they'll go. If your steering wheel is adjustable, move it fully in, making sure the steering wheel is straight (wonky steering wheels look sloppy in photos). Close the glove box, remove the hanging air freshener and the key out of the ignition. Check the armrest lid and all the door bin lids are down and close the cupholders.

Together with the sparkly clean glass, these tweaks help create an illusion of space and light in your interior, which will translate to the images you capture. Next, get snapping, not forgetting to change the angles and height of your shooting as you go, and do capture those gleaming detailed pedals as well! Anyone who details their pedals is a proper Driveway Warrior, right?

Clear Up

Here we go again with the boring but still important bit. The idea is to do it in such a way that all your kit is instantly ready to go for your next detailing session. Start by returning all your liquids, and spray bottles to storage. Do first check that all lids and spray heads are secure and keep an eye on any dilutables that need topping up. If you used the Handheld Foam Sprayer, rinse it out with clean water, semi-pressurise it, and spray out some clean water to clear the feed pipe and nozzle. Give it a quick dry and store it away.

Grab your Vacuum Cleaner and extension lead, Caddy, Kneeling Mats and Stool, give them a quick wipe down if necessary and store them away. Gather your used Cloths and pop them in the washing machine at 40°C (104°F), add the Non-Bio Delicates detergent and stick on a 30-minute cycle with a gentle spin. When they come out of the machine, give them a good shake to reset the fibres and hang everything on a clothes dryer to dry naturally.

You should be left with your various Brushes and Magic Eraser. Dump them all in your clean Wheel Bucket, add a small squirt of washing-up liquid to the Bucket, and half-fill it with hot tap water swirling it around to mix. Washing-up liquid is an excellent degreaser, perfect for cleaning in this way. Spend a few minutes with your gloved hands in the soapy water, massaging the Brushes clean.

Then dump the soapy water and half-fill the Bucket again with clean cold water. Give the Brushes and Eraser a bit of a massage in the water to get the soap suds out, and then dump the rinse water in your water butt if you have one (your garden won't mind the little bit of soap).

I then give the Magic Eraser a squeeze and all the Brushes a good wrist flick to fling off some of the excess rinse water. If possible, put everything in the sun to dry, or failing that, lay them on an old dry towel under the radiator to dry. Give your Wheel Bucket a quick wipe to dry and store it away. The next day when your Cloths, Magic Eraser and Brushes are fully dry, store them away, keeping your Wheel Brush in your Wheel Wash Bucket.

Reflection

So, now your interior matches the quality of your exterior. Your cockpit is transformed into a tranquil, harmonious place in which to pilot some seriously fun, spirited road testing. Or, kick back and snag those icy beers while you reflect contentedly on the £195 ($263) you've saved on a professional interior detail job. Not to mention the substantial future savings your newly acquired detailing skills, and detailing kit, will afford you.

Read on in Chapter 3 to find out what happened when a dirty great rock assaulted my car and chipped the paintwork and how I fixed it, Driveway Warrior fashion …

CHAPTER 3
Stunning Stone Chip Repair

I was cruising down the motorway in the Cayman at a steady 70mph, without a care in the world, listening contentedly to the deep, *resonant* growl of the flat-six power plant just behind me. A sharp CRACK just in front of me rudely shattered the tranquillity of the moment.

"What the hell?!" I exclaimed to no one in particular, instinctively ducking my head.

Oh crap, I thought, I bet I know what that ferocious assault on my car was!

As I pulled into the next pit stop to investigate the source of the brutal, shatter-er of peace and harmony, my worst fears were confirmed. Right on the passenger side, there was a *filthy* great stone chip right on the front of the roofline. Just where the top of the windscreen meets the roof.

"Oh, you @$*!%#!" I scolded the offending crater.

Now, where did I put my paint stick?

The truth is you won't notice when 99% of stone chips happen to your motor. However, unless you stick to 40mph, don't ever drive within 20 metres behind another vehicle or have Paint Protect Film (PPF) fitted – (I do use PPF now), the destructive little buggers will appear with inevitable *stealth* and *cunning* accuracy. They will particularly seem to target your front bumper, bonnet (hood), and rear flared wheel arches.

How to Get the Correct Colour Paint Stick

Thank goodness you have allies in your car's paint code and a generic paint stick! Your local Official Porsche Centre (OPC) will happily relieve you of £20

($26.84) for their weeny Original Equipment (OE) paint stick, as will any other car dealership. But why pay through the nose when you can easily ask Mr Google where your paint code is located? It is often in your service book or on a sticker somewhere in your boot (trunk). Then, use the paint code to buy a generic paint stick from eBay for a fraction of the OE price. Go for a kit that includes paint and clear coat lacquer sticks for an enduring two-stage fix.

Colour Matching

The Aqua Blue (paint code LM5R) paint stick I bought for my Cayman from eBay for £6 ($8.05) is a perfect colour match, and my front bumper and roof look much smarter as a result! Yours can as well, and if you have concerns over potential colour matching, no problem. Take your paint stick, shake *very* well, coat a small square of plain white paper, and allow it to dry. Then, in good light, use auto detailing tape to fix your paint card to your bodywork; stand back and see what your eyes tell you. If you're not happy with the match, there's no harm done, and most suppliers will refund you for unsatisfactory colour matching if you ask nicely!

Identifying the Stone Chip

You might find it helpful to examine the area around your stone chip to see if you can identify any other chips that need attention. That way, you can get them all done while you have your kit prepared and you're in the chip *zone*! Get yourself a super bright LED light to help with this; they're dirt cheap online; my fearsomely bright generic branded, palm-size light was £5 ($6.71) from eBay, and I've had it three years.

When Should I Fix the Stone Chip?

Stone chip repair isn't only about aesthetics; it also protects your car from corrosion. If the stone chip is bad and causes exposed primer, or worse, exposed bare metal, the *dreaded* rust can quickly take hold and spread if you don't fix it. So, you need to act fast to preserve and protect your precious paintwork. Fortunately, it's pretty easy to do with just a few tools and a steady hand - if you know how. As a heads up, to do the job properly, you need to wait 48 hours between Stage 5 Clear Coat application and Stage 6 Paint Refinement to let the paint and clear coat cure fully.

What's the Weather Like?

Again, with the weather conditions! All the advice given in previous chapters about detailing in ideal weather conditions apply to your stone chip repair project. May I refer you to Chapter 1, *Sunshine – the Safe Wash Enemy*.' Never start or finish your Detailing Processes in direct sunlight if you want to avoid issues with your detailing liquids drying out too quickly on your car and screwing up the quality of your finish. Pick a day when the forecast is dry and mild, preferably overcast, when your paintwork is cool to the touch. We want to give ourselves every opportunity to achieve the perfect results, right?

Impeccable Paint Chip Repair Kit List

Detailing Hardware Required:

- Carry Caddy
- Nitrile Gloves
- Cotton Buds (Q-tips)
- Clean Microfibre Cloths
- Cheap Artist Brush Set
- Kneeling Mats
- Folding Stool
- Container for neat IPA
- Hairdryer
- Small piece of Clingfilm (Saran Wrap)
- Piece of Kitchen Roll

Detailing Liquids Needed:

- Paint Stick
- Clear Coat Lacquer Stick
- Washing-Up Liquid (Dish Soap) Soapy Water Spray Bottle
- All-Purpose Cleaner (APC) Spray 1:5
- Isopropyl Alcohol (IPA) neat 99.9%
- Scratch Remover *or* Cutting Compound
- Super Resin Polish
- Spray Wax

- Microfibre Detergent
- Solvent Cleaner to clean the paintbrush

Top Tip Washing-up liquid is an excellent degreaser and is perfect for the initial cleaning of the stone chip area. Take one of your spray bottles, half fill with warm water, add a squirt of Washing-up liquid and shake well – bingo, a dirt-cheap turbocharged degreaser!

Preparation

You know the drill by now; remember to get those beers in the fridge, ready for when you finish, and grab a bottle of water or Thermocafé mug of coffee to sustain you. Remove your rings and watch, and make sure you aren't wearing anything zippy or with buttons or poppers that could scratch your car while you're working up close to it. Position your car preferably with room to safely work around it. Put your Paint and Clear Coat Sticks indoors to get to room temperature; they don't perform well when too cold or hot. Put your Kneeling Mats and Folding Stool in position if you're working down low.

Get your Stone Chip Repair Kit assembled and load your Carry Caddy with your Spray Potions, Microfibre Cloths, Cotton Buds (Q-tips), Artist Brushes, etc. Have your Caddy handy.

5 Preparation Stages

Stage 1

Washing-up Liquid Spray Down: With your Gloves on, take your Washing-up Liquid Spray Bottle and spritz down the stone chip area, extending the spray out on all sides of the chip. Then, grab a clean Microfibre Cloth and dry off the whole area.

Stage 2

APC Spray Down: Next, grab your APC Spray Bottle and spray everything down with APC to remove the salt content of the Washing-up Liquid from your paintwork. Dry the APC off with a clean Microfibre Cloth.

Stage 3

Scratch Remover or Compound: You might not be able to see them, but your stone chip(s) will have tiny, jagged, raised paint edges caused by the impact of the stone *smacking* into the paintwork. Ideally, these raised edges should be wet sanded if the stone chip is large. But I don't cover bodywork wet sanding in this book as I understand it may be a *scary* concept for Driveway Warriors! No problem. There's a great compromise for most stone chips – Scratch Remover or Cutting Compound. The safe abrasive materials within both will mimic the outcome of wet sanding, though in a more controlled and subtle way.

> ***Top Tip*** Remove the lid from the Cutting Compound or Scratch Remover bottle, and if they are mucky on the inside from previous use, thoroughly rinse them under a hot tap. You don't want to transfer any dried-out, powdery compound or polish residue onto your cloth and then onto your car. Dry the lids and screw them securely back on the bottles – it's all in the detailing!

Grab your Scratch Remover or Cutting Compound; whichever you have will work fine and shake well. Take a clean Microfibre Cloth, double it, hook it over your fingertip, apply a small blob about half the size of a pea to the Cloth, and work it over the stone chip. Go at it with medium pressure using *short* strokes in different directions for 30 seconds, staying just outside the chip area. This will help to level down those invisible miniature raised edges, creating a relatively level zone that will help give your Paint Stick something on which to bond. Finish this stage by buffing off the area with a clean Microfibre Cloth.

Stage 4

IPA Cleansing: This is the final cleaning stage before breaking out your Paint and Clear Coat Sticks. Grab your source bottle of neat IPA Liquid 99.9% and pour a *small* amount of the neat IPA into your container. Then, dip a Cotton Bud (Q-tip) in the IPA and, holding the Cotton Bud down near the tip for ease of control, use it to thoroughly clean in and around the stone chip area.

You have now fully prepared the stone chip to receive paint and then clear coat. The neat IPA will finish cleaning the chip thoroughly and promote the adhesion of the touch-up paint and clear coat. Used in this way, the IPA will *flash* away quickly, and you shouldn't need to wipe it down with a cloth.

Stage 5

2-Step Paint & Clear Coat Application: Take your Paint Stick, which should be at room temperature by now, and give it a *thorough* shake for at least two minutes; this is very important and set it aside. Pick out a nice *fine-tipped* brush from your cheap Artist Brush Set. Give the bristles a gentle pinch and pull between thumb and forefinger to make sure the bristles won't shed in your perfectly applied new paint!

2 Paint Steps

Step 1: Base Paint First Coat: Remove the supplied brush from the Paint Stick and put it somewhere for safekeeping. You don't want to use it as you will usually find this brush is a bit rubbish, e.g., the bristles are too long or thick, and when you take it out of the container, the paint drips down the brush stem, creating the potential for a hideous mess – nightmare! Dip your Fine-Tipped Paintbrush into the paint so you have a *small* amount just on the tip of the Brush. If you get too much paint on the Brush, blot it gently on a piece of kitchen towel.

Then, with a steady hand, holding the Paintbrush approximately 2cm (1in) from the base of the bristles, slowly and gently dab the paint into the chip until you have the whole chip covered in paint. What you're aiming for here is to fill the chip with paint but *not* have the paint sitting *above* the surface of the chip. So, go easy and make sure you don't overload your brush with paint.

It's best to do this in two separate coats, as a proportion of the paint comprises paint thinner, and you may find the first coat will *sink* a bit when the paint thinner has evaporated after approximately 15 minutes or so. Wrap the Brush tip tightly in Clingfilm (Saran Wrap) so the bristles don't dry out. You can use the Hairdryer on the paint between coats to speed up the drying.

Base Paint Second Coat: Allow the initial base paint coat to dry for approximately 15 minutes, then carefully apply a second coat. Remember just to fill the chip and *not* have the paint sitting *above* the surface of the chip. Once again, wrap the brush tip tightly in Clingfilm (Saran Wrap), so the bristles don't dry out until you have time to clean the brush later.

Step 2: Clear Lacquer First Coat: When the base paint is dry to the touch, it's time to apply the Clear Coat over the Base Paint. Dip a new, fine-tipped Artist Brush into the Clear Coat so you have a *small* amount just on the tip of the Brush. If you get too much lacquer on the Brush, blot it gently on a piece of kitchen towel.

Then, with a steady hand, holding the Paintbrush approximately 2cm (1in) from the base of the bristles, slowly and gently dab the Clear Coat into the chip until you have the whole chip covered in Clear Coat. Wrap the Brush tip tightly in Clingfilm (Saran Wrap). Use the Hairdryer on the Clear Coat between coats to speed up the drying.

Clear Lacquer Second Coat: In the same way that the base paint did, once again, the Clear Coat will likely slightly sink as it dries. When the first Clear Coat is dry to the touch, go ahead and apply a second Clear Coat over the chip.

Leave It to Dry for 48 Hours

You now need to exercise *superhuman* levels of patience and allow the clear coat to cure for 48 hours fully. If you don't allow it time to harden, the paint will come off when you attempt to refine the finish, wash, wax, or polish the car, and you'll have to start again – *noo*! However, it's fine to drive the car after an hour or so when the Clear Coat is touch dry.

Clear Up

Aah, the inevitable joys of clear up, gird your loins, and here we jolly well go again – at least this one is a quickie, hallelujah!

The idea is to do it in such a way that all your kit is instantly ready to go for your next detailing session. Start by returning all your Liquids, Paint, and Clear Coat Sticks and Spray Bottles to storage, but do first check all lids and spray heads are secure and keep an eye on any dilutables that need topping up. There's nothing worse than going for your detailing spray, only to find there's only a dribble left!

Grab your Kneeling Mats and Stool, give them a quick wipe down if necessary and store them away. Gather your used Cloths and pop them in the washing machine at 40°C (104°F), add the Non-Bio Delicates Detergent and stick on a 30-minute cycle with a gentle spin. When they come out of the machine, give them a good shake to re-set the fibres and hang everything on a clothes dryer to dry naturally. Clean the Fine-Tip Paint Brushes thoroughly with solvent cleaner and store them away. The next day when your Cloths are fully dry, store them away safely.

Stage 6

48 Hours Later - Stone Chip Paint Refinement

Scratch Remover or Compound: Ok, so you've *heroically* managed to leave your repair alone, unmolested for 48 hours, and you can now refine the repair finish – who knew there was so much flipping work to fix a stone chip? It's all in the detailing, my friend. This step is similar to Stage 3, and you're looking to level down the protruding clear coat layer to be flush with the surrounding paintwork. So, retrieve your Scratch Remover or Cutting Compound - whichever you have will work fine - and shake well.

Take a clean Microfibre Cloth, double it, and hook it over your fingertip. Then apply a small blob about half the size of a pea to the Cloth and work it over the stone chip. Go at it with medium pressure using *short* strokes in different directions for 30 seconds, staying just outside the chip area. This will help level down any protruding clear coat raised edges, making them flush with the surrounding paint. Finish this stage by buffing it off with a clean Microfibre Cloth.

Stage 7

Final Polish: You're at the final polishing stage. This stage will restore gloss to your repair area, so grab your Super Resin Polish and shake well. Take a clean Microfibre Cloth, double it, hook it over your fingertip, then apply a pea size polish blob to the Cloth and work it over the stone chip. Go at it with medium pressure, this time using longer strokes and going in different directions, so you extend out, working the polish over the surrounding paintwork.

Extend the time you work this polish over the paint to 45 seconds, allow it to dry for a further minute, and buff it off with a clean Microfibre Cloth. Call me fussy, but I like to finish off by *locking in* the gloss with a quick squirt of spray wax and a final buff with a clean Microfibre Cloth. You are DONE! Phew, congratulations, Driveway Warrior, on a job *superbly* done!

Reflection

So, no Stone Chip repair is ever going to be perfect without a full, flipping expensive respray. But I'll happily wager that what you've achieved in this project at *minimal* cost is a hundred times improved from the filthy great offensive paint chip you've banished! Be proud of that, my friend; you are *awesome!*

It's difficult to price what professionals charge for stone chip repair, as it depends on the severity and number of chips in question. One thing is for sure, it *won't* be cheap, and I have seen off-the-shelf touch-up kits for £70 ($95) that *don't* incorporate half the bits and pieces we used in this project. So, I'll leave *you* to judge the value of this project, my friend, though I think I know which way you'll go.

Read on in the next chapter for my 'Desirable Detailing Products & Tools' review to help you get *stunning* professional standard detailing results every time in your future Driveway Warrior adventures …

Chapter 4

Desirable DIY Detailing Products & Tools

Part 1 - Detailing Hardware

Ever wondered how pro detailers produce such *stunning* results? Sure, they have tried and trusted techniques, but it's also down to the perfect detailing products at their disposal. With the *dizzying* array of detailing products available on the market, it can be daunting to decide what to start with. In this chapter, I take all the guesswork out of which detailing products are best for you; I exclusively reveal how to build your basic killer detailing kit and the approximate cost. I take you through the industry's secret hacks and techniques to achieve massive savings on the high-quality products you should use on your Porsche or any sports car.

My wife calls my Cayman a *'menoporsche!'* Your ride is your baby, right? So, she deserves the best treatment, and you can quickly build your detailing kit for a modest sum, with a bit of 'know-how' in your up-front investment! Using the best products will instantly help you achieve fantastic, dependable results every time. They will also make the process easier for you while ensuring the results endure and stand the test of time.

If you spend several hours detailing your ride, you want the results to last as long as possible, right? These products will help you get the job done effectively with a durable finish and make you proud of the fabulously self-detailed beast you have unleashed.

Safe Wash Kit

Perhaps the key element of your home detailing set-up will be your Safe Wash Kit (I explain about 'Safe Wash' in Chapter 1). Get some of these items added to your Christmas and birthday wish lists! I just love them because they're superb in value and performance.

I've split the items into categories, e.g., shampoos, polishes, sealants, towels, cloths, etc., and divided them again into two sections: 'hardware' and 'liquids,' for your ease of reference.

Detailing Hardware:

Pressure Washer and Hosepipe: If you're serious about detailing your pride and joy, get yourself a reliable Pressure Washer. There are some brilliant bargain deals out there these days; my favourite brand is Danish company Nilfisk. I bought a budget *Nilfisk C110* model well over ten years ago; it has been super robust and dependable and just keeps going, despite receiving regular driveway abuse! Ensure any machine you buy comes with the 'wide fan' spray nozzle for use on cars, or you risk damaging your paint. It also needs a dedicated snow-foam dispensing bottle. If you really can't run to a Pressure Washer, a Hosepipe with an excellent adjustable Nozzle is essential. I prefer *Hozelock* Hoses and Nozzles for value and performance.

Vacuum Cleaner: For my wet & dry Vacuum Cleaner, I wanted something lightweight, manoeuvrable, compact, powerful, and keenly priced, not much to ask, right? My workhorse of choice is the *'Nilfisk Buddy II 12,'* 12-litre model, which has proved the perfect choice for tackling mucky interiors. The 1200-watt motor is more than powerful enough to get your interior spick n span, and it has a blower function for another cleaning option.

The built-in accessory holder handily stores a good selection of tools, and it's surprisingly robust, stable and easy to move around. The compact design makes it easy to store, and best of all, its superb value online for approximately £50 ($66.92) is well under half the price of a similarly specced Henry model!

Buckets and Grit Guards

Wash and Rinse Buckets: You'll need three buckets and three grit guards to safe wash. I use my two trusty, dedicated wash and rinse 20-litre buckets with label stickers on the side. Each bucket has a lid to keep them clean in storage, and they also need a grit -guard, a plastic grill designed to sit at the bottom of the buckets.

It is ultra-effective in trapping the grit and keeping it away from your wash mitt. These can be found inexpensively online for approx. £15-£25 ($20-$33). Pick wash & rinse buckets with different coloured stickers for wash & rinse.

The third bucket is for your wheels. I use a basic 10-litre builder's bucket, cheap as chips and highly robust. My dedicated wheel wash bucket is a different colour from my two safe-wash buckets, is clearly labelled "Wheel Wash," and I added a third grit guard to it. DIY stores are the best place to pick up a builder's bucket for approximately £2-£4 ($2.68-$5.35).

Mitts

Wash Mitts: Never use scratchy sponges, an old t-shirt or chamois leather, etc., to safe wash your car if you don't want to introduce swirls and scratches all over your paint! Instead, get yourself two cheap paint-friendly super-soft wash mitts in different colours – one for your paintwork and one for your wheel bucket. My favourite is the *'Kent Car Care Microfibre Noodle Wash Mitt'*. These are brilliant for gently lifting dirt away from your car and holding the dirt below the noodles' surface before dunking it in your rinse bucket. Again, these are colossal value and super durable; I recently picked up two online for just £5 ($6.72).

Clay Mitt: Having safe washed and dried your Porsche, if the paint still feels slightly rough to the touch, it will need manual decontamination claying as part of your 'Paint Correction' process. Once upon a time, the prospect of claying struck fear in the hearts of home detailers. Those pesky, pricey little lumps of car clay that went rock hard in the cold and had to be binned the first time you dropped it.

However, salvation has arrived in the form of the splendid *'G3 Professional Deep Clean Clay Mitt,'* which has revolutionised how we can clay our cars; I explain in detail how to use it in Chapter 8. Used correctly, this mitt will safely remove bonded contaminants, ingrained dirt and tar spots. They are widely available for approximately £12 ($16).

Drying Aids

Drying Towel Extra Large: Having safe washed and air blasted rinse water from the nooks & crannies with your Car Blower, the next job is to safely contact dry your car using a dedicated Microfibre Drying Towel. Look no further than the immense in size and value, *'Kent Car Care Extra Large Microfibre Drying Towel'*. This vast yellow towel measures in at a mighty 80 x 62cm (31.5 x 24.5in). It's lint-free and clear coat safe, satin-edged for safety and durability, and positively guzzles rinse water by absorbing eight times its weight in water from large areas fast. Don't forget to rip off the label, and similar to the Halfords Microfibre Cloths, this one works best when slightly damp. The great thing is it's less than half the price of many of its rivals. When finished, pop in the washing machine at 40°C (104°F), line dry thoroughly and store in a clean bag, and it will work perfectly for you time after time. They are widely available for approximately £7.50 ($9.99).

Car Blower: Now I'm aware there are various brands of dedicated vehicle blow dryers out there, most of which cost well over £100 ($134), and some also require an air compressor. I realise this may be a controversial solution among some pro-detailers, but what I use for contactless drying cost me £15 ($20) new from eBay. It's simply a cheap, unbranded Chinese-made mains-powered mini leaf-blower, measuring 38cm x 16cm (15in x 6.5in), and it's perfect for blasting rinse water out of all your car's nooks, crannies and crevices before towel drying. It has been seemingly indestructible throughout the four years I've been using it, as well as being a rock-solid, reliable performer.

Granted, this blower doesn't expel warm air, but it's more than powerful enough for the job and completely fit for purpose. I've also used this blower successfully to help me fully re-furbish two sets of alloy wheels at home (think blowing off endless sanding dust). It's a brilliant, cheap solution, and the compact size makes it a cinch to store.

You could even use it for leaf-blowing when not detailing your car! These can be found in different colours on Amazon and eBay for approximately £15-£30 ($20-$40). You just need to add a decent extension cable.

Microfibre Cloths: Most of us know these are super soft cloths, perfect for all manner of detailing tasks. But not all Microfibre Cloths are made equal!

You'll be amazed by how many cloths you get through in a detailing session, but which ones should you confidently go for? I've been impressed by the quality, value, and reliability of *'Halfords Microfibre Cloths'* 40 x 34cm (16in x 13.5in). These can be used on all your interior and exterior surfaces; for washing, polishing, waxing, glass and upholstery cleaning, etc. The polyester/polyamide composition is highly absorbent, and they work best when slightly damp as this aids dirt removal. They will wick away water while reducing friction on your paint, thus minimising the risk of introducing swirl marks.

The packs come with different colour cloths, which enable you to avoid cross-contamination by separating your interior and exterior cloths. When finished detailing, give them a good rinse and pop in the washing machine at 40°C (104°F). As your older cloths get grubby, relegate them to wheel and engine use, and your Microfibre Cloth supply will give you years of service - perfect! They are available for £3.99 ($5.32) for a pack of 5 or £6.99 ($9.32) for a pack of 10.

> *Top Tip* Always rip the label off the cloth before use; believe it or not, it can cause tiny scratches when dragged over the paint!

Buffing Towel: Once you've applied your polish, wax, or ceramic coating of choice, it's time to buff your paintwork to a glorious, glossy shine. My go-to is the *'Glart Premium Soft Microfibre Towel,'* 40 x 40cm (16 x 16in). It's anthracite in colour and has a plush, deep pile, fleece texture, perfect for polishing and buffing paintwork.

This is the softest towel I've ever used; the polyester/polyamide composition feels more like a luxurious bath towel than something you use on your car. The slick red satin edging assists safety and durability. Don't forget to rip off the label, and when finished, pop it in the washing machine at 40°C (104°F). Line dry it thoroughly, store the towel in a clean bag, and it will loyally serve you again and again. They are available online for approximately £14 ($18.66) for a pack of 3.

Microfibre Applicator Pads: Applying wax, polish, or UNO Protect One Step to your paintwork is a breeze with these lovely soft, round, blue pads. They have a finger pocket on one side, big enough even for sausage-sized fingers! If you're polishing the whole car, it's advisable to keep a good supply of these to hand, so you can change pads when they become polish-clogged. After use, give the pads a quick scrub with an old toothbrush, blot with a Microfibre Cloth, and pop them in the washing machine at 40°C (104°F). Dry them naturally, store them in a plastic bag, and they'll keep on giving for you. They are available online for approximately £9.60 ($12.79) for a pack of 10.

Brushes

Wheel Cleaning Set: I now take you through my ultimate tried and tested Wheel Cleaning Tools Kit. I've lost count of the number of different shapes, sizes, and textures of wheel brushes I've tried over the years. I've finally settled contentedly on just five dedicated wheel cleaning tools that get the job done perfectly - every time!

1. 'Meguiar's Supreme Wheel Brush Medium 37cm (14.5in).' This has a devastatingly soft, plush, fluffy microfibre head perfect for alloys and long enough to reach through to the back of the wheel. The head holds a ton of wash water, providing epic lubricity, and importantly this size is just right to squeeze past most spoke and caliper designs.

It's nice and robust; after a year's hard use, it's still going strong with no sign of damage. It has no wire or metal parts to damage your delicate alloys, and the sturdy plastic handle is foam cushioned for comfort and grip. It is available online for approximately £14 ($18.67).

2. 'ValetPro Chemical Resistant Wheel Brush (Plastic Handle).' Go in with this sturdy, supple-headed brush after the *Meguiar's Supreme* to eradicate any remaining areas of grime. It has durable though soft, cool-looking blue bristles to agitate the muck away gently. It is available online for approximately £8 ($10.67) or cost-effectively as part of the detailing brushes kit described in the next section.

3. 'Vikan Detail Brush Set 30mm & 50mm.' These two long-handled, soft-bristled brushes are angle headed, making them the perfect size and shape to reach through your spokes to give your brake calipers a thorough satisfying cleansing. They are available online for approximately £7 ($9.34).

4. 'Kent Car Care Microfibre Noodle Wash Mitt.' I describe this earlier in the chapter. Use your dedicated wheel bucket wash mitt to reach around the back of the spokes to target this often-forgotten blind spot. The back of the mitt has a textured surface ideal for agitating behind the spokes. If you don't, you'll always know the hidden grime lurks there, mocking you!

5. 'in2Detailing Tyre Brush.' This is a brilliant, cheap, short-handled, stiff-bristled, lightweight tyre brush (never use this on your wheels as it will scratch them). It's ergonomically designed to deep clean tyres - keep it well lubricated with wash water, spray your tyre walls down with APC, and get scrubbing! You'll have your tyres clean and ready to receive tyre dressing in no time. It's available online for a blinking snip at approximately £3.50 ($4.67).

Detailing Brushes Kit (interior & exterior): Detailing brushes are an essential addition to your kit to enable you to gently agitate and remove stubborn grime as part of your contact safe wash process. I wanted an economically priced, quality brush kit that would cover interior and exterior detailing tasks, and I found fulfilment in the *'ValetPro Detailing Brushes Kit.'* This is a set of three plastic-handled brushes of different sizes and textures. One of which I keep in my wheel cleaning bucket (covered in the previous section).

The other two brushes comprise a 'dash brush,' with a soft pure boar's hair head that's brilliant for dusting tricky to reach areas around interior switches, vents, seams and knobs! The third is a 'large sash brush,' with a soft-bristled 1-inch head. Keep it super-well-lubricated with wash water, and it's perfect for agitating muck around exterior paintwork areas like badges, grills, lights, and body trim after you've snow foamed your car.

After use, clean thoroughly with hot water and washing-up liquid, dry naturally before storing, and they'll perform unendingly for you. The brush kit is available online at approximately £12.60 ($16.80).

Upholstery and Leather Care Brushes: I round off the brushes section with two inexpensive additions to your kit, an upholstery brush and a leather care brush.

Firstly, we have the *'Trade Quality Large Upholstery Brush.'* You'll spot this wallet-friendly, solidly built, distinctive yellow and blue bad-boy among your detailing supplies a mile off! The stiff bristles are specially designed to make short work of dirt and grime on your mats and upholstery. It has a 'grippy' ergonomic handle and is used with your handheld foam dispenser. It is available online from £2.86 ($3.81).

Next up is the *'Gyeon Q2M Leather Care Brush'*. This cute wee brush has a pleasing wooden, ergonomically shaped handle that moulds nicely into the hand, giving a decent amount of grip even when damp. Your delicate leather seats will thank you for favouring them with gentle horse-hair bristles, designed to be as safe and effective as possible on all leather types. This super little brush is available online for approximately £3.50 ($4.65).

Handheld Spray Bottles & Foam Dispenser

Handheld Foam Sprayer: This little handheld beauty is a formidable ally in your detailing arsenal. The *'IK 1.5 Foam Sprayer'* provides a dense, durable foam that's perfect for tackling your filthy wheels, wheel arches, engine and bodywork paint, etc., as a pre-wash. Yes, it turns your detailing liquids into thick, lasting foam! Simply fill the wide neck bottle to the 0.75-litre mark with a 1:5 mix of (All-Purpose Cleaner to de-ionised water), pump the handle to pressurise, and unleash the foam over your mucky wheels. Then allow to soak briefly and complete the job with your wheel bucket kit.

The great thing about this sprayer is that it truly does economise your detailing liquids by stretching the cleaning power of your APC, Snow Foam, etc., bringing out their maximum potential. It is brilliant, not to mention a piece of cake and oh so satisfying to use. It even comes with different spray nozzle options. I've found that one full load is enough to foam up all four wheels and wheel arches or to cover the entire lower third of your car. It is available online for approximately £17 ($22.73).

Detailing Spray Bottles: Hopefully, by now, you'll have an insight into the economising power of the ability to stretch your detailing liquids to their maximum potential by diluting them with de-ionised water. To do this, you'll need a decent supply of half a dozen or so sturdy spray bottles for your various detailing liquids. Look no further than the *'Discount Cleaning Supplies Colour Head Spray Bottle 750ml'*. They have a chemical-resistant, robust feel with a good size trigger and an adjustable nozzle. The broad base means the bottle won't easily tip over on the go in your Carry Caddy and the imprinted side measuring gauge is extremely handy for diluting your potions. The other cool element is the ability to buy spare parts from the company, e.g., a replacement colour head spray trigger is just £0.88 ($1.17)! You can even colour code to help avoid cross-contamination; just label the bottle with a permanent marker; what could be easier? They are available online in assorted colours for £1.46 ($1.95) per bottle.

Handy Hardware:

Tyre Hose Guides/Hose Jam Eliminator: "What the hell is that?" I hear you cry! If you've ever suffered from your hose or power cable getting caught under your tyres, slowing you down and driving you nuts, TRUST me; you need these genius little gadgets from *'Detail Guardz'* in your arsenal. Simply place each Guide under the outer edge of each tyre, and your hose or cable will glide gracefully across the inbuilt roller, speeding you happily toward a stress-free job well done. Simple but brilliant and well worth the money. They are available online in different colours for approximately £11 ($14.72) as a 2-pack or approximately £15.95 ($21.35) as a 4-pack. I have found the 2-pack to be enough for my Cayman and VW Golf.

Carry Caddy: When you're detailing around your car, you're guaranteed to work faster and with increased convenience by having your detailing products handily at your side. I use the sturdy *'Wham Carry Caddy'*; it has plenty of storage space for your selected kit, plus a large, well-balanced centre handle. It comes in different colours (yes, mine does match my Porsche), is simple and foolproof to use! Load up your caddy with your selected sprays, cloths, etc., and you'll wonder how you ever coped without one! They are available online for approximately £4 ($5.33).

Trim Removal Tool: You're probably wondering why I've included this one in a detailing kit list. Well, it's because they're fantastic for specific tasks when a detail brush just won't cut it. Think of the inevitable, irksome 'bird bombs' delivered with stunning precision, penetrating deep down between the panel gaps on your pride and joy or where the rubber trim joins the paintwork. It happens with infuriating frequency and is potentially a nightmare to remove, so what to do? Simply select your cheapo appropriately-shaped flat-ended plastic trim tool, take your clean Microfibre Cloth, wrap it over the trim tool and soak the end with APC: you have the perfect implement to prise out the unwelcome deposit! I have a basic 5-piece set that cost £6 ($8) from Amazon, and it can even be used for trim removal! I cover some of these projects in later chapters.

Magic Eraser: Cast aside your scepticism and open your mind to perhaps the most surprisingly clever and definitely cheapest detailing hack on this list. The *'Flash Magic Eraser Ultra Power.'* You'll find these either in your kitchen cupboards or the cleaning section at your local supermarket. Ever wondered how to tackle all those annoying scuffs and scratch marks dragging down your interior's aesthetics? Here's the solution. Take your Magic Eraser, moisten it with APC and work it using medium pressure over the interior scuff in a crosshatch pattern, keeping the eraser lubricated with APC. You'll be amazed at the restorative effect. Then just finish off your fix with a squirt of dressing, worked in with a clean MF cloth. This works particularly well on those textured interior panels and surfaces where a Microfibre Cloth may be less effective. They are available online for approximately £2.60 ($3.47) for a pack of 2.

> ***Top Tip*** This technique also works brilliantly on exterior paint 'transfer' scuffs; you just need to be careful to focus the eraser on the 'transfer' mark as much as possible and not on the paintwork. I think you'll be chuffed by how efficiently this technique lifts the 'transfer' mark. Finish off your paint transfer fix by buffing the area with your favourite polish, followed by a squirt of spray wax.

Personal Protective Equipment (PPE)

Kneeling Mats: Time for some practical PPE. The last thing you need when you're detailing down low is fractured kneecaps. Call me a wimp, but I prefer as much comfort as possible in life and in detailing! This is where the *'Halfords 6-piece Black Floor Mat Set'* comes to your rescue. This 6-piece shock-absorbing, interlocking foam mat set is perfect for reducing discomfort on those kneecap-busting wheel cleaning or lower third paint polishing sessions.

They're easy to clean and water-resistant for ease of maintenance. Each mat measures 60 x 60cm (24 x 24in), more than large enough for your knee's gratification. I like to place a mat on all four sides of the car as I move around detailing down low, for which my knees always thank me! A cool aspect of the 6-piece set is the ability to keep 4 for your knees and the other 2 mats to use as part of your car winterising/storage process. I will cover this in the next book! This set is a steal at approximately £12 ($16).

Folding Stool: While we're on the subject of detailing in comfort, it would be rude not to mention the *'Knight Plastic Folding Stool/Step.'* At H29 x L27 x W22cm (11.5 x 10.5 x 8.5in), this mini stool is great for all those down low or up high detailing jobs and is excellent for prolonged, seated, mucky wheel cleaning sessions. It's made of heavy-duty, easy-clean plastic for durability and has a high grip surface, suitable for 100kg (16st). It has the bonus of being lightweight, portable and space-saving due to the cool fold-flat design. Some detailers use wheeled stools, but those are no good for high-up work and are useless on a sloping driveway. It is available online in assorted colours for approximately £11.90 ($15.88).

Gloves: Think safety first and cover up with *'Black Nitrile Gloves.'* These premium industrial-grade gloves are 50% thicker than standard nitrile gloves, with a far greater puncture resistance than latex. While great for your car, many detailing products contain chemicals that won't do your skin any favours! A home detailer's must-have, I've found I can re-use these multiple times, which means a box will last years. They are available in all sizes online from approximately £15 ($20) for a box of 100.

Misc. Detailing Bits 'n' Pieces

I include here a few handy items that don't really need a dedicated section but will nonetheless be helpful in your detailing escapades at some point:

Cotton Buds/Q-tips: These are splendid for cleaning stone chips in your paint ahead of using your paint stick repair. Simply dip the end in IPA and work over the stone chip; it will flash off in no time, leaving the surface ready for you to do the Paint Stick touch-up.

Window Blade/Squeegee: After your safe wash final rinse, use a squeegee to slide the water down off your windows (don't use a squeegee on your paintwork as it may scratch). You don't need anything special; just go for something comfy to hold with a nice flexible blade for a couple of quid.

Funnel Set: You'll find a funnel set super helpful for organising your dilutables into your detailing spray bottles. They are available online as a 4-piece set measuring 50-120mm, for peanuts, from £1.63 ($2.15).

Soft Toothbrushes: These help clean the worst of the polish residue from applicator pads before the pads go in the washing machine. Child toothbrushes are the softest and cost a few pence from supermarkets.

Artist Brush Set: These are a lifesaver when it comes to stone chip touching up! With a Paint Stick, you usually find the brush is a bit rubbish, and when you take it out of the container, the paint drips down the brush stem, creating the potential for a hideous mess. I bought my Artist Value Brush Pack from eBay for £3 ($3.99); it has ten assorted brush head sizes giving a perfect selection for stone chip repair.

Peek Premium Metal Polish: If your chrome exhaust tips have lost their lustre, Peek does a superb job of bringing them back to life. I cover this in chapter 6.

Grabber/Reacher Tool: This long reach tool is perfect for retrieving anything dropped into hard to reached spots, e.g., when detailing your engine bay. They are available in different lengths, and you definitely need one of these in your detailing kit.

LED Light: My fearsomely bright generic branded, palm-size light was £5 ($6.79) from eBay, and I've had it for three years and counting. You'll need one of these for some of your projects, e.g., stone chip repair, paint correction, and undercarriage detailing to name a few. They are super bright and dirt cheap online.

Wheel Chocks: These wheel blockers are essential safety items for use with ramps and other lifting devices. I prefer the heavy-duty, seemingly indestructible *Rhino Gear chocks* at around £11.99 ($16.23) for a pair online.

WD-40 Spray: Surely, the famed enemy of moisture requires no introduction! It has endless uses in detailing, and I find the 450ml cans with the inbuilt applicator straw most helpful.

Part 2 - Detailing Liquids

Detailing Liquids

Rust Repellent: Have you ever noticed how your steel brake discs (rotors) turn that nasty orange colour after you safe wash your ride? Well, it's basically the formation of pesky surface rust, and it totally ruins the look of your freshly detailed car. You can get rid of the dreaded orange hue by driving your car and firmly hitting the brakes a few times; this will scrub the surface rust away. But I also have a brilliant little iron-clad (excuse the pun) hack to prevent the evil orange foe.

I use *'Rust Destructor,'* a powerful and safe rust repellent/dissolving liquid. The beauty of this stuff is that it can be economically diluted down, it's non-toxic, does not harm good steel, paint, plastics, or rubber, and it's biodegradable! Simply mix with de-ionised water to make the required quantity; I've found that a 1:3 mix (Rust Destructor to water) works nicely. Decant to a spray bottle and give each disc a good spray down before safe washing your car. The spray solution will form a transparent, flexible barrier to rust formation, like magic – job done! This is available online in a 500ml bottle for approximately £8.95 ($11.98) or 1 litre for approximately £14.95 ($20). When diluted down with a 1:3 ratio, it will last for ages.

Cleaning Liquids

Snow Foam: Undoubtedly, Snow Foam is one of the most satisfying and pleasurable products to play with (I mean, safe wash your pride and joy with). My go-to product is the luxurious suds of *'Autoglym Polar Blast,'* a pH-neutral solution that will not strip existing layers of polish or wax. It works as a pre-wash by creating a rich, clingy foam blanket that grabs the entire car surface. The contact time enables it to break down dirt and grime, lifting it off your car's surface. In turn, this makes your safe wash steps much faster and more effective, reducing the risk of scratches and swirls caused by your wash mitt moving over dirt.

You'll need a Pressure Washer to get the best snow foam. Dilute down 1:5 (Snow Foam to Tap Water) in your Snow Foam Sprayer; no need for de-ionised water here, and away you go! A word of warning: Autoglym advises it isn't suitable to use on soft-top convertible roofs.

I love the dilutable element of this Snow Foam, making it a solid, economical keeper in my Safe Wash Kit. It is available online from £13 ($17.36) for 2.5 litres.

Detailing Solution Mix Water: A staple for the home detailer, use *'De-ionised Water'* to safely dilute your detailing products to the required level; any brand will do. De-ionised water is also used in car batteries and steam irons. It's cheaper than dirt and available in supermarkets and motoring accessory outlets for approximately £1.50 ($2) for 5 litres. Never use tap water as it will leave behind water spots and residue that you will then have to wash off the car.

All-Purpose Cleaner (APC): This is another must-have for the home detailer. I've trialled several brands, and my champion of choice is the tried and trusted *'CarPro Multi X.'*

It has tremendous, safe, flexible cleaning power, suitable for interior and exterior use; its concentrated formula ruthlessly tackles all levels of dirt and contamination on your vehicle. What I love about it is the ability to dilute it down. Mix it with de-ionised water, and it will just keep on going for you. It has fantastic versatility; decant to a sprayer and dilute it 1:20 (Multi X to Water) for general cleaning or as a bodywork pre-wash, 1:50 for a gentle interior cleaner, or 1:5 for mucky floor mats. I also add 2 capfuls to my safe wash and wheel wash buckets for guaranteed additional cleaning and foaming power. It is available online in 500ml bottles for approximately £11 ($14.70), 1 Litre for about £19 ($25.42) or 4 Litres for around £48 ($64.21). For context, my 1 Litre bottle is still going strong after a year.

Glass Cleaner: I've trialled different glass cleaners, and my favourite is *'Stoner Invisible Glass'.* I like the comparatively large bottle size (650ml) and the fact it's safe to use inside and out on tinted windows found in many sports cars. The chunky feel trigger is nice and easy to use, and it's not grabby to buff off with a clean Microfibre Cloth. It contains no soap or surfactants and flashes off quickly, cleaning thoroughly without leaving any streaks, haze, or residue. It's just the ticket for gleaming glass and is available online, costing approximately £9 ($12) for 650ml.

Waterless Wash (WW): I include this cracking, though cheapo, detailing spray for one reason. I present you with one of the lowest cost products in this book, *'Triplewax Waterless Wash.'* Nature forces me to use it most days – it's brilliant at conveniently removing 'bird bombs' from your pride and joy's paint or glass! When the winged terrors deposit their pay-load over your gleaming ride, as soon as possible, you need to take your WW, shake well and liberally douse both the pay-load and a clean MF Cloth with it. Allow it to soak for a minute, then carefully wipe the offending item away, avoiding rubbing and circular motions. The shampoo and wax formula flashes away nicely, does not streak, and leaves behind an impressive waxy shine. A great, economical solution for an inevitable nuisance! It is available from £4.50 ($6) for a generous 1-litre spray bottle.

Isopropanol Alcohol 99.9% (IPA): No, you can't take a sneaky swig to armour you on your detailing adventures! However, it needs to be an essential element of your detailing kit if you want professional results. You won't usually find IPA on the shelves of your local auto accessories store, more likely in the chemist (pharmacist). IPA is a vastly versatile and super economical liquid for detailing. Buy the 99.9% proof stuff, and it can be diluted down to 70% proof with de-ionised water for all your detailing needs. It's a cinch to use as it flashes rapidly away. Simply mix 70:30 IPA to de-ionised water in your 1-litre spray bottle, shake well, and away you go. When diluted in this way, it's super safe on your paintwork.

Use it as a paint correction panel wipe to cleanse paintwork before compounding/polishing and as a paint cleanser to prepare for ceramic coating or paint stick stone chip touching-up. It's also phenomenal as a plastics and rubbers prep before applying dressings, as it enables the dressing to adhere fully and with lasting results. I also used IPA to help me re-furbish my alloy wheels in prepping the surface prior to successfully re-spraying. Various brands and sizes are available online; they all do the same thing. Go for the cost-effective, long-lasting 99.9% proof 5-litre container for approximately £19 ($25.41).

Microfibre Detergent: When it comes to laundering your hard-working microfibre cloths, mitts and soft buffing towels, it's essential to use a gentle detergent that will clean effectively without damaging the fibres in your wash media. You can easily spend a wallet load on branded detailing Microfibre Detergent, but there's really no need. I use the un-sexy, though effective *'Asda Non-Bio Delicates Wash.'* It will look after your microfibre cloths, towels, and mitts, ensuring you can use them time and again. I tend to separate my interior and exterior cloths before they go into the washing machine.

When cloths get very mucky, chuck them in a bucket half-filled with warm water, add a small squirt of washing-up liquid and give them a short pre-wash with your gloved hand. Then squeeze them out and pop them in the washing machine at 40°C (104°F) on a 30-minute cycle with a gentle spin. The only thing I wash separately is my Glart Soft Buffing Towel; it's a bit more delicate, so I do it at 30°C (86°F). When you take them out of the washing machine, give them a good shake to re-set the fibres. Hang everything on a clothes dryer to dry naturally, using pegs for the mitts. It is available for £2 ($2.65) for 630ml.

Upholstery & Carpet Protector

Fabric Protector: Some of us have experienced that moment of panic as your cappuccino-to-go slips agonisingly from your grasp, all over your spotless carpet mats. Fear not; you can pre-protect your carpets and mats, making them substantially easier to clean in the event of a disaster striking. My favourite product is *'Scotchgard Multi-Purpose Protector'*; it forms an invisible barrier and will protect your leather and fabric from water, greasy stains and oil-based liquids. Just give your mats and carpets a good vacuum, spot clean with APC foam if necessary, and then evenly spray on a thin coat in a crosshatch pattern. I usually leave it to cure for a few hours before using the car. It's great value at approximately £8 ($10.66) for 400ml and well worth adding to your detailing kit.

Bodywork Shampoos

Bodywork Shampoo: When it comes to washing your car, never use washing-up liquid (except in one particular circumstance I explain in Chapter 8) as it contains salt that will rust your car over time. I've tried several car shampoos, and by far, my favourite shampoo for performance and value is the excellent *'Autoglym Bodywork Shampoo Conditioner.'* It's a divine smelling pH neutral shampoo (meaning it won't strip previously applied polishes/waxes and won't dry out your trim). It's also outstanding hydrophobicity if you want to be technical! The low foaming solution contains a water repellent film that helps rinse water to glide satisfyingly off the surface. It's an absolute doddle to use and does not streak.

Available in 500ml for approximately £6.50 ($8.69) and 1 litre for around £10 ($13.36). However, the one to go for is the bargain 2.5 litre for about £17 ($22.82) online. Use 2 capfuls in your wash bucket, and 2.5 litres will last ages!

Bodywork Shampoo (Wax and Polish Stripping): You will only need to use this as part of your 'Paint Correction process.' I cover this in Chapter 8. *'Autoglym Pure Shampoo'* will dependably strip previously applied layers of wax and polish; the only time you want to do this is in preparation for ceramic sealing your paint. 'Pure' is a safe, easy-to-use, high foaming shampoo that cleans thoroughly without streaking. It contains no wax or gloss enhancers, which leaves nothing behind, ideal for paint correction prep. I know some home detailers use washing-up liquid for this step, though if, like me, you can't quite bring yourself to drench your car in rust-causing destructive salt-laden dish soap, then 'Pure' is my alternative. I only buy this in small quantities to avoid waste. It is widely available for approximately £5.80 ($7.74) for 500ml or £7.85 ($10.47) for 1 litre.

Dressing Solution

Dressing Solution: Without a doubt, my absolute favourite detailing product is the magnificent *'CarPro Perl'* water-based silicone-oxide all-purpose Dressing Solution from South Korean company CarPro. It's the only dressing you'll ever need!

"Why magnificent?" I hear you ask.

Well, it's just so astoundingly flexible, with a seemingly endless list of ways to use it on the interior, exterior, and engine with stunning results. 'Perl' is an acronym for 'Plastic, Engine, Rubber Leather,' and the fantastic thing is it can be diluted down with de-ionised water in a spray bottle at different strengths, depending on the required task.

I use it on my exterior rubbers and plastics diluted at a ratio of 1:3 (Perl to water); on my tyre walls diluted at 1:2; as an engine dressing at 1:2; and as an interior trim and leather dressing diluted at 1:5, to name but a few uses. The restorative effect is like magic, darkening trim and bringing a subtle mid-satin sheen without an overly vulgar glossy finish. An absolute must-have for the home detailer, it is available in 500ml for approximately £11.50 ($15.41) or better still, 1 Litre costs around £18 ($24.13) and will last for ages when diluted down.

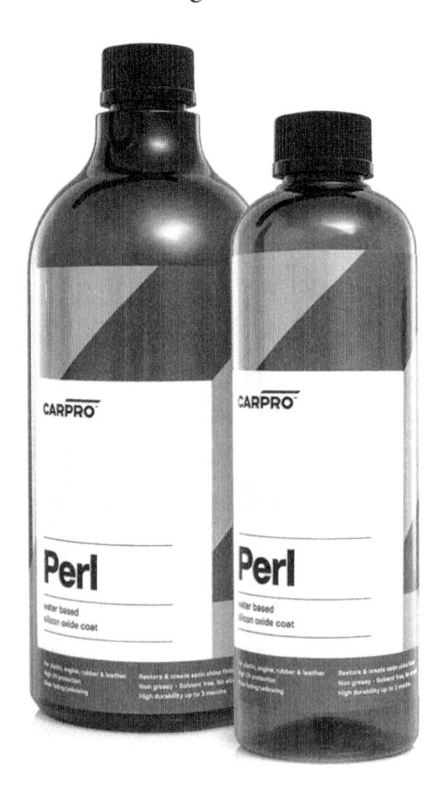

Chemical Decontamination

Fallout Remover: This stuff is formulated to remove iron contamination particles that have bonded to your paintwork or wheels. It will form part of a 'Paint Correction' process at the 'chemical decontamination' stage, which I cover in Chapter 8. A word of warning: use your face mask as all fallout removers reek of sulphur! My favoured foul-fragranced brand is *'Autobrite Direct Purple Rain Iron Fallout Remover,'* as this one is less offensive on the nostrils than most! The pH-balanced formula it's safe to use on your paintwork, fast-acting, and effectively dissolves iron contaminants and industrial fallout. Just hold your nose if you have delicate nostrils and observe the immensely satisfying sight of the purple colour reaction as it gets to work. If you haven't passed out by the time you've finished using it, job's a good 'un! It is available online for approximately £8.10 ($10.80) for a 500ml spray bottle.

Tar and Glue Remover: If your car has tar spots over the lower third, you're going to need a tar remover that's friendly but effective on all the exterior parts. *'Autosmart Tardis Powerful Solvent Cleaner'* is super safe on your paintwork and plastics and will rapidly remove tree sap, tar and glue deposits, wax, oil and grease. It can be used as a spot cleaner or decanted to a spray bottle and used during 'paint correction.' It is important to note that this one is not dilutable. It is available from the Autosmart website for £20.90 ($27.60) for 1 litre or on eBay for £27.69 ($36.56) for 5 litres.

If you only need a tiny amount of tar remover, *'Autoglym Intensive Tar Remover'* will safely do the same job, available online from £6.49 ($8.57) for 325ml.

Polishes, Cutting Compound, Wax, Sealants & Scratch Remover

Cutting Compound: We have here another legendary performer, this time in the business of body paint compounding. Put simply, a cutting compound is an abrasive material suspended in a paste, used to restore car paintwork. My choice is *'Meguiar's Ultimate Compound.'* Its advantage is that it's clear coat safe and cuts as fast as harsh abrasives without scratching or hazing. It authentically does restore colour clarity to faded and neglected finishes. I used it as a 2-stage compound/polish treatment on the 'tired' paint front bumper on my Cayman. The results were so pleasing, I was able to abandon my plan to have the bumper expensively repainted by a body shop – result! It can be used with a hand applicator or electric polisher; the flip-top cap makes it easy to control how much liquid solution you dispense, and it's excellent for removing stubborn 'bird bomb' residue. It is widely available online for approximately £15 ($20) for 450ml.

Polish: As all car enthusiasts know, maintaining the beauty and quality of your paintwork is vital to your happiness. For me, there is only one hero of choice, the legendary *'Autoglym Super Resin Polish.'* This polish has been around forever and for a good reason. Used correctly, it gives flipping magnificent results, imparting a deep, rich glossy finish. You can use it with a hand applicator or budget machine polisher (I have done both). It works on all paint types, and the gentle abrasive properties effectively remove surface contaminants and oxidisation left behind from your safe wash. It also does a fantastic job masking light swirls and imperfections in your paint. Less is more with this polish, so the key is to apply it sparingly to avoid 'dusting' up. I explain in detail how to use it in Chapter 8. The larger bottles are great value and are widely available in various sizes: 325ml, 500ml, and 1 litre, for between approximately £8.95 to £13 ($11.95 to $17.35).

One-Step Polish & Sealant: Now, I totally get that not everyone has the time or the inclination to perform a complete DIY 7-Stage Paint Correction procedure. Fortunately, there's a super alternative that comes pretty darn close to matching the gloss and shine of a 7-stage process, if not the level of protection. *'RUPES UNO PROTECT One Step Polish & Sealant'* offers you that alternative. It's a true 'all-in-one' compound, polish and protectant product, capable of removing light defects and producing a stunning high gloss mirror finish on your paintwork in just one step! It uses a rich blend of polymer, silicone and carnauba to put down a durable, protective layer capable of lasting up to three months.

The cool thing is that you can either apply it by hand with a microfibre applicator pad or with a budget machine polisher. I have tried both methods with favourable results.

After the safe wash, dry the car, shake the UNO well, apply a small amount to your microfibre applicator pad, and work on the paint with moderate pressure. Allow a minute for the UNO to 'set up' before buffing off with your dedicated soft buffing towel. I used it as a 2-stage compound/polish treatment to the tired-looking front bumper on my Cayman. The results were so pleasing that I was able to abandon my plan to have the bumper expensively repainted by a body shop. Thank you, RUPES UNO! It is available online for approximately £9.95 ($13.25) for 250ml or £28.75 ($38.28) for 1 litre.

Wax: Again, I've trialled loads of different paint bodywork waxes, many of which, though effective, can be a massive pain to use due to the insane effort and time required to buff the stuff off. However, I've eventually landed on *'Autoglym Rapid Aqua Wax.'* It is a heavenly-scented carnauba-based liquid spray wax, offering unrivalled ease of use, combined with stunning guaranteed results every time. Simply safe wash your car, then do 2-3 sprays of Wax per panel onto the still wet paintwork. Spread it with a clean Microfibre Cloth, then buff it to a glossy shine with a separate clean, Soft Buffing Towel.

It is a fantastic product, available online as a 500ml spray bottle for approximately £13 ($17.47), and this also comes with a decent quality Microfibre spreading Cloth! Better still, go for the heavily discounted value of the *'Autoglym Express Wax'* 5-litre container for around £34 ($45.71). This is the same product as the 500ml spray bottle, just re-branded for commercial use. Shake well, decant into a spray bottle, and it will last you for years. I've found the 500ml spray bottle lasts about six car washes, the only place I don't use it is on the glass.

Ceramic Spray Coating: This is the BIG DADDY - essentially science in a spray bottle! Until fairly recently, ceramic coatings seemed to be the secret preserve of the pro detailing industry, and the relatively high product price was an obstacle to home detailers. However, all that has changed. Ceramic coating technology has advanced leaps and bounds, and several products have come onto the market in recent years, becoming accessible and affordable to all – hallelujah! The product I use on my Porsche to remarkable effect is *'Turtle Wax's Hybrid Solutions Ceramic Spray Coating.'* I cover how to use this in detail in Chapter 8.

This coating really will help your car stay clean for longer. It's formulated with SiO2 and delivers phenomenal hydrophobicity plus 12 months of ceramic protection when used correctly. The synthetic wax polymers will increase your car's depth of colour and gloss and give a brilliant mirror-like shine. The exquisite fruity fragrance is rather a treat on the nostrils as well! Available from approximately £14 ($18.62) for 500ml, I've found this will make three separate applications on my Porsche (six coats in all).

Alloys, Brake Calipers & Exhaust Tip Ceramic Coating: If you want to give your expensive alloy wheels the love and care they deserve, you must ceramic coat them. This is where the diminutive but formidable bottles of *'Gtechniq's C5 Wheel Armour'* come to your service. C5 uses a scientific formula to chemically bond the ceramic coating to your rims, making them repellent to brake dust and contamination. This makes washing them oh so much easier; a blast with the pressure washer is usually sufficient to make them gleam.

I cover the detailed wheel protection process in Chapter 6. You'll need to get your rims off the car to decontaminate them properly. Prep them with IPA, apply a few drops of C5 to the mini-applicator pad, and spread methodically over your rim front and back. Follow the same process to protect your calipers and exhaust tip(s), though these can stay on the car! Available online for approximately £22.80

($30.37) for 15ml, or £41.80 ($55.69) for 30ml, it's not cheap, but it's flipping good. In my experience, one 15ml bottle will do four alloy wheels front and back. The other advantage is that a single C5 application done correctly gives up to 12 months of protection.

Scratch Remover: Working around your Porsche doing a safe wash, ideally during the contact drying towel stage when you're up close and personal with the paintwork, is the time to keep a keen eye out for the dreaded minor scratches and scuffs. Like 'bird bombs,' they are an inevitable nuisance for any car driven spiritedly or parked on public roads. By keeping on top of these unsightly blemishes, you preserve the beauty of the paint; you'll enjoy your car all the more and ultimately help keep the value up. My trusted choice is *'Farécla's G3 Permanent Professional Scratch Remover.'*

I like this formula because it doesn't just mask or 'fill' the scratch; it uses engineered diminishing abrasives to remove it permanently. This will only work on minor clear coat scratches that haven't penetrated the paint. An easy way to check this is to spray the scratch with water; if the scratch temporarily disappears, it's a clear coat scratch, and you can tackle it. To use, clean around the scratch with IPA, lightly dampen a clean Microfibre Cloth with water, apply a pea size blob of G3, and using a medium, even pressure, go over the scratch in a crosshatch pattern. Then take your Super Resin Polish and buff with a clean Microfibre Cloth, and bingo, stand back and admire your efforts with pride.

Cracking value, available at approximately £9.50 ($12.64) for 500ml, it will last for donkey's years!

Stone Chip Repair

Paint Stick: Sadly, your fast B-road weapon can be a magnet for the inevitable stone chips, especially on the front bumper and flared rear wheel arches. However, you have allies in your car's paint code and a generic Paint Stick!

Your local Official Porsche Centre (OPC) will happily relieve you of £20 ($26.59) for their weenie OE Paint Stick. But why pay through the nose when you can easily ask Mr Google where your paint code is located (often in your service book) and then use it to buy a generic Paint Stick from eBay for a fraction of the OPC price? The Aqua Blue (paint code LM5R) Paint Stick I bought for my Cayman from eBay for £6 ($7.95) is a perfect colour match, and my front bumper looks much smarter as a result. I explain colour matching in Chapter 3.

Reflection

No doubt there's a lot of kit to digest here, not to mention to potentially fork out your hard-earned cash on. No one will rush out and order everything on the list straight away, and neither should they do so. First, I would suggest browsing through all the detailing projects covered in this book. Then perhaps cherry-pick your initial favoured project(s), and using the specific project 'Kit List,' obtain the hardware and liquids kit you need, gradually building up your gear in this way.

If you're not sure where to get started, I suggest a good Safe Wash (Chapter 1) would be a good starting point to give you a relatively swift bang for buck result on the exterior of your ride. Then, follow that up separately with an Interior Detailing session (Chapter 2). Go for it, Driveway Detailing Warrior; you can do this …

CHAPTER 5

Wicked Wheels off Wheel-Arch Detail & Protection

Beginning your detailing journey feels great, and life is good. You'll have transformed your ride into a thing of beauty, the envy of all who behold! It's only fitting that you turn your attention to the remaining visible exterior area letting the side down. As you gaze lovingly at your gorgeous alloys, perhaps with the sun glinting alluringly off the angular, sculpted alloy perfection, maybe your eye is drawn to the murky backdrop of wheel arch framing your handsome rims in a less than flattering way. The grime is lurking there, seemingly out of reach, mocking you - it will always be there, nagging away at the back of your mind until you take action to *banish* it.

Despair not, for you now have the detailing skills, know-how and kit to *oust* those drab, grubby wheel arches and transform them into a gleaming backdrop befitting of a worthy frame for your sexy rims.

"How so?" I hear you ask.

Well, that would be with a Wicked Wheels off Wheel Arch Detail and Protection project. Let's get cracking!

Weather Conditions

May I refer you back to the section in Chapter 1, *'Sunshine – the Safe Wash Enemy.'* All the advice given in previous chapters about detailing in ideal weather conditions apply to your Wheels off Wheel Arch Detail Project. Never start or finish your Detailing Processes in direct sunlight if you want to avoid issues with your detailing liquids drying out too quickly on your car and screwing up the quality of your finish.

Pick a couple of days when the forecast is dry and mild, preferably overcast when your paintwork is cool to the touch. We want to give ourselves every opportunity to achieve the perfect results, right?

Wicked Wheels off Wheel-Arch Detailing Kit List

Detailing Hardware Required:

- Carry Caddy
- Nitrile Gloves
- Kneeling Mats
- Folding Stool
- Pressure Washer Kit & Extension Lead *or* Hosepipe and Nozzle
- Wheel Wash Bucket & Kit
- Car Blower
- Detail Guardz
- Handheld Foam Sprayer (diluted 1:5 APC: De-ionised Water if not using a Pressure Washer)
- Microfibre Cloths (a bag full of downgraded cloths)
- Trim Removal Tools
- Eye Protection
- Face Mask
- Scotch Pad
- Container for Wheel Nuts

Detailing Liquids Needed:

- Rust Repellent (diluted 1:3)
- Pure Shampoo
- Snow Foam (diluted 1:4 tap water)
- Perl Dressing Solution Spray (diluted 1:3)
- All-Purpose Cleaner (APC) Spray (diluted 1:5)
- Isopropyl Alcohol (IPA) Spray (diluted 70:30)
- WD-40
- Fallout Remover Spray
- Tar & Glue Remover Spray
- Non-Bio Microfibre Detergent

Inevitably, you'll need to remove your wheels, either one or two at a time, depending on what kit you have. No problem, though the procedure for lifting a 1430 kg (3153 lbs) vehicle can be a little daunting for the Driveway Warrior, particularly the first time. However, with the right tools and observing the basic safety steps covered in the following pages, safe wheel removal is a basic procedure you CAN tackle. I have done so dozens of times on my driveway, and you can do it as well.

Wheel Removal & Replacement Tools:

- 2 Sets of Axle Stands (Jack Stands), i.e., 4 in total *or* a Rennstand Safe Jack
- Low Profile Hydraulic Trolley Jack
- Wheel Chocks x 4
- Breaker Bar & *Wheel Nut Socket
- Torque Wrench
- Wheel Hanger Tool
- Locking Wheel Nut Key
- Container for Wheel Nuts
- Copper Grease
- Aluminium Paste

Most Porsche wheel nuts/bolts are 19mm; check the handbook for your correct size.

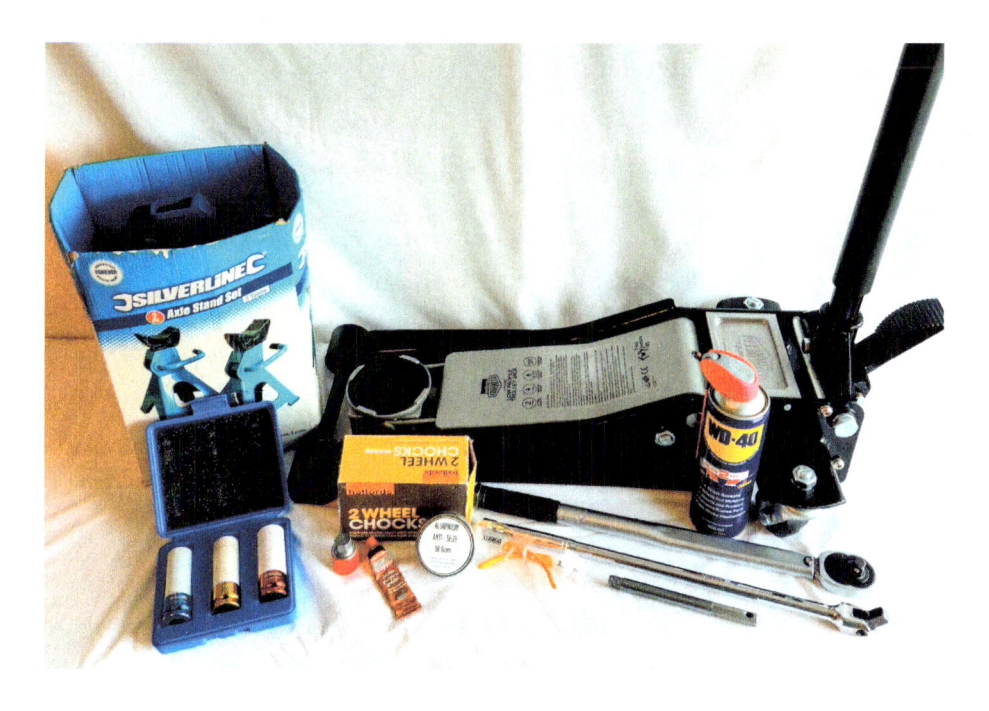

Top Tips Use the plastic-sheathed variety of wheel nut socket head when removing and replacing your wheels. The plastic coating over the socket head will help prevent damage to your precious rims. Wrap some insulation tape around the head of your locking wheel nut key to help prevent the key from scuffing against the inside of your wheels while working with the locking key.

Preparation

Remember to get those beers in the fridge, ready for when you finish, grab a bottle of water or Thermocafé mug of coffee to sustain you.

Remove your rings and watch, and make sure you aren't wearing anything zippy or with buttons or poppers that could scratch your car while you're working up close to it. Position your vehicle on a level, flat, stable surface *(please no sloping driveways for this project)*, preferably with room to safely work around it. Place your Detail Guardz under the outer edges of the tyres (of the wheels not being removed). Get your Wheels off Wheel Arch detailing kit and wheel removal/replacement tools assembled. Load up your Carry Caddy with your Spray Potions, Microfibre Cloths, Wheel Hanger Tool, Breaker Bar, Wheel Nut Socket, Locking Wheel Nut Key, etc., and have your Caddy handy. Put your Kneeling Mats and Stool in position but out of the way of your Trolley Jack.

A Word on Low Profile Hydraulic Trolley Jacks

I use the *Halfords Advanced 2 Tonne Low Profile Hydraulic Trolley Jack* to lift my Cayman. It has a substantial, robust feel, and it meets all my needs. The minimum height of only 75mm (2.95in) easily fits under my low Cayman, and it lifts to a height of 505mm (19.9in). The dual piston quick lift foot pedal speeds things up nicely, and the removable jack pad helps prevent damage to your undercarriage. I like the nice big jack saddle diameter of 110mm (4.3in) with a rubber pad which provides an excellent contact area. Best of all, Halfords give a 3-year warranty – *brilliant!* It is available from Halfords for approximately £132 ($176).

I *do not* recommend the smaller style of mini trolley jack for this task, and I regard these mini jacks as suitable for *emergency* use only. The reason being the jack *saddle* tends to be much smaller, and therefore its contact area under the lifting point is somewhat limited. Years ago, I attempted to lift my car using one of these mini trolley jacks, and I experienced a *scary* moment when the jack saddle partially slipped off the jacking point. Fortunately, disaster was averted, and I learned an important lesson about always using the correct tools.

> **Top Tip** If you're unfamiliar with your Trolley Jack, I highly recommend you thoroughly read the instruction booklet and practice with the jack away from the car before you try and lift the vehicle. Raise and lower the jack saddle to get a feel for how it works, does it release slowly or quickly, etc. Naturally, I don't condone childish antics such as standing on the jack saddle and jacking yourself up for a laugh!

5 Steps to Safely Jacking Up Your Car & Removing the Front Wheels

Make no mistake, incorrect use of a Hydraulic Trolley Jack can lead to expensive damage to your pride and joy, or worse, injury to yourself. The alternative is to pay a pro detailer around £240 ($321) to do the project for you. But if you have the kit, a safe area to work, and the confidence to have a go, let's *DO* this.

Before You Raise the Car

Step 1: I repeat, make sure the car is on a flat, level, stable surface; once again, please *no* sloping driveways for this project. Remove any excess weight from the vehicle, e.g., spare wheel or children! Locate the locking wheel nut key, remove the ignition key and firmly apply the handbrake. If you put the car in park (for an automatic gearbox/transmission), it will only lock the two rear wheels. Check that your Axle Stands are all in *good* condition. If anything looks badly corroded, bent, or damaged, please *bin* it immediately; it simply isn't worth the risk to use any defective kit!

Chock the Wheels & Loosen the Nuts

Step 2: Put your Gloves on and carefully chock the front and back, both sides of each wheel, for all the wheels not being removed; this is an extra precaution against the car rolling. So, if you're raising the front of the vehicle to remove the front two wheels, you need to chock both sides of the same two rear wheels (two chocks per wheel). The reason for doing this is that most surfaces are not completely level – as always, safety first, friends!

While the wheels are still on the ground, use the Breaker Bar and Wheel Nut Socket to 'crack' open (loosen one turn) all 5-wheel nuts, working anti-clockwise on each front wheel. If you don't do this now, your wheels will spin, and you'll have a headache trying to loosen the wheel nuts. Don't forget you'll need the Locking Wheel Nut Key for your locking wheel nuts. Most cars have one locking wheel nut on each wheel.

Warning *Never, under any circumstances, work on a raised car* using *the* trolley jack alone, *without axle stands being in place to support the car's weight. Not even if it's just for a minute. Also, be aware that the handbrake will no longer work when you raise the rear of the car (as it only works on the two rear wheels).*

Raising the Front of the Car

Step 3: This is the longest step as there's a fair bit of safety stuff to take care of to ensure Driveway Warriors will battle on another day! Consult your owner's handbook to identify the car's four factory standard jacking points. These are the strengthened sections on the chassis where it's safe to put either your jack saddle or Axle Stand. However, the obstacle here is that there isn't enough room to fit your trolley jack lifting saddle and your Axle Stand together under the same jacking point.

This problem is eliminated if you're lucky enough to own a *Rennstand Safe Jack*, a great bit of kit that enables safe lifting *and* supporting together at the same correct spot. Unfortunately, at the time of writing, my understanding is that the *Rennstand* is still only available in the USA. You can order one direct from Safe Jack in the USA, though you'll have to fork out for shipping wherever in the world you are. If memory serves me correctly, I paid about £70 ($93) for shipping my Rennstand to the UK - well worth it, in my opinion.

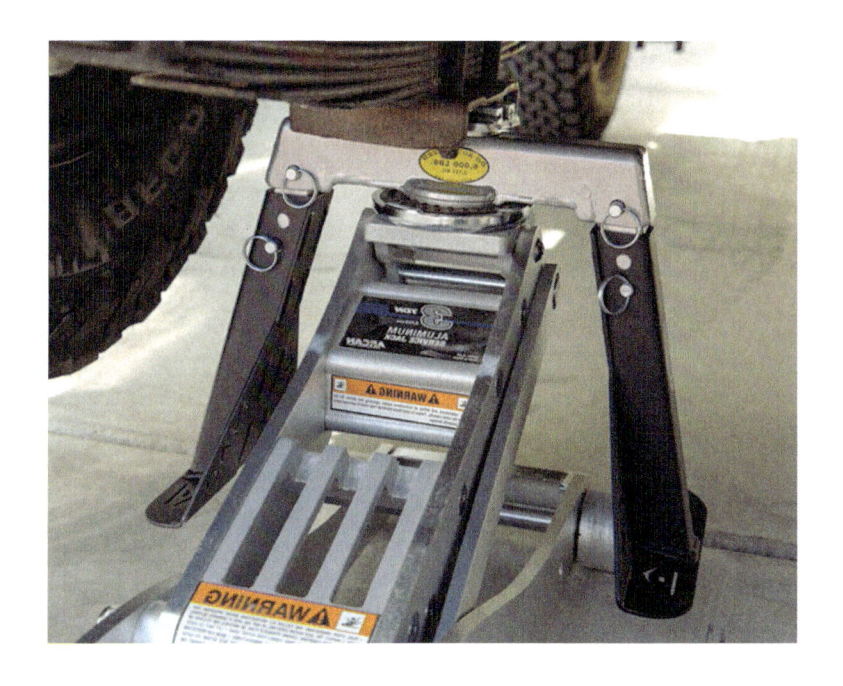

So, how to get around this challenge without a Rennstand? I prefer to lift the front end of the car first, and I only remove the front two wheels before *replacing* them on the vehicle and *lowering the front* of the car before going on to raise the rear of the car. The car is far more stable, with two wheels remaining on the ground. My Cayman has a reinforced area of the front chassis, just behind the front factory standard jacking point, which is the ideal point to jack the car up - your car should have a similar area. If you place your Trolley Jack under this section to lift, you'll have enough room to fit your Axle Stands under the factory standard jacking point at the same time. Don't place the Axle Stands under any part of the engine or gearbox, as it may make the car unstable.

Take your trolley jack and line up the jack saddle underneath the reinforced area of the front chassis, just behind the factory standard jacking point. I like to place a folded Microfibre Cloth on the jack saddle to prevent metal on metal contact from scraping against the undercarriage.

Have your Axle Stand ready beside you, check that it's set at the correct height, and the adjustable height pin is pushed firmly home. Raise the jack saddle slowly, checking it's lined up perfectly under the reinforced area.

Lift the car slowly until the wheel clears the ground by approximately 2cm (1in).

It's normal for the car to tilt slightly while the wheels on the opposite side remain on the ground. You only need to lift the vehicle enough to enable you to slide the Axle Stand set to its minimum height (if it's high enough to hold the car off the ground) securely under the factory standard jacking point. Carefully line it up under the factory standard jacking point. Once the axle stand is in position, carefully and slowly lower the car onto the Axle Stand and remove the Trolley Jack away from the car out of your way.

> ***Top Tip*** At this point, I like to put in another safety precaution. Place emergency third and fourth Axle Stands *loosely* underneath the lifting points on each side of the car that your trolley jack has just vacated. These emergency Axle Stands are your backup in the unlikely event of your main Axle Stand failing.

Repeat the same process on the opposite side of the car, so you have both front wheels off the ground and supported with Axle Stands. Ensure both Axle Stands are set to the same height so that the car sits level and secure.

Warning Never put your fingers between the car's jacking point and the axle stand during the lifting process.

Note: *When you start to lift the car, it's important to be aware that depending on where you place your trolley jack saddle, it's possible both wheels on the same side of the car could rise off the ground, or both front wheels may raise off the ground at the same time. However, only the wheel nearest the jack will rise if you use the above-recommended lifting point.*

Check the Car's Stability

Step 4: We need to do one final safety check before we remove the front wheels.

Now, gird your loins, gather your courage and faith, and standing at the raised end, give the car a firm shove to check if it's super stable on the Axle Stands.

If the car doesn't budge, you're good to go. But if the car moves at all, you do *not* have it correctly supported. It's far better that the vehicle falls off the Axle Stands while you're shoving it from the side than falling on any part of you while you're working on it!

Wheel Removal

Step 5: Finally, it's safe to remove the front wheels – phew! So, you've already loosened the wheel nuts during Step 2. Grab your Breaker Bar, Locking Wheel Nut Key, Wheel Nut Socket and the Container for Wheel Nuts, and further loosen all five wheel nuts on one wheel.

Be careful not to do what I did once and hit the handle of the breaker bar against your bodywork – ouch! When you've got the nuts loosened to a certain point, you might find it easier to take the wheel nut socket off the breaker bar and finish removing the nuts by holding the socket directly in your hand.

Top Tip First, remove the wheel nut closest to 12 o'clock, then screw your Wheel Hanger Tool (mine is from *Bloxsport*) securely through the vacant wheel nut opening into the wheel nut hole on the wheel hub. Finish removing the four other wheel nuts and put all five nuts safely in your Container. Grasp the wheel firmly with two hands and give it a hefty wiggle. Depending on how professional the last person to replace the wheel was, it will easily come away from the hub, *or* it may resist and require a bit of wrestling. When the wheel comes free of the hub, it will drop *blissfully* onto the wheel hanger and not crash sickeningly down onto your painted brake caliper, causing a horrendous chip on the caliper in the process!

Slide the wheel off the wheel hanger and slip the wheel underneath the sill near the primary Axle Stand as a final safety measure if anything falls. Repeat the same process on the opposite side. Move your Lifting Gear and Wheel Removal Tools away from the car somewhere safe, out of the way.

8 Stages to Wicked Wheel-Arch Detailing and Protection

Grab your Caddy loaded with your detailing products, and let's get into some actual detailing! Take your Wheel Bucket and add two capfuls of Pure Shampoo plus two capfuls of APC for extra foaming and cleaning power. Fill with warm water and swirl the water to mix the shampoo and APC. Make sure your Grit Guard is in the bottom of the Bucket.

Stage 1

Rust Repellent & Initial Rinse Down: Gloves on, and as you'll primarily be working down low with the wheel arch at eye level, please put your eye protection on as you don't want any of these chemicals splashing in your eyes. Grab the Rust Repellent Spray, shake well, and give each brake disc (rotor) a good spray down. Allow to cure for two minutes, and move your Rust Repellent away from the car, out of the way.

Give a thorough rinse down under the arches with your pressure washer or hose nozzle. Ensure your pressure washer attachment is the wide fan nozzle and keep the powerful spray moving to loosen and dislodge the worst of the muck under there. Make sure you get the bit of the arch that sits directly behind the lip roll of your wheel arch. You can't see it without contorting your head under there, but grime tends to build up there, and over time it will rust your arches through unless you tackle it. Reach under that point with your gloved fingers, and you'll feel the build-up of debris under there; you need to *banish* it!

Stage 2

Pre-Soak with Snow Foam or APC Foam: If using a Pressure Washer, generously coat under the wheel arch with Snow Foam 1:4 mix. If not using a Pressure Washer, load up your Handheld Foam Sprayer with a 1:5 mix (APC to Water), shake well, and generously foam up everywhere under the wheel arch. Allow the Snow Foam or APC Foam to soak for two minutes, then give the whole area under the arch a second thorough rinse down with your Pressure Washer or Hose Nozzle.

Stage 3

Brush Attack with Snow Foam or APC Foam: Now, for the second time, generously coat under the wheel arch with Snow Foam or Handheld Sprayer APC Foam and allow to sit for two minutes. Then take your Wheel Wash Bucket & Kit and, using your Kneeling Mats and Stool if you wish, get scrubbing everywhere under the arch.

Start with the Meguiar's Supreme Brush, regularly rinsing it in your Wheel Wash Bucket and work your way through your Wheel Brush Kit. Keep them super-well-lubricated with wash water as you go (a dry brush is a scratchy one).

Use the Wheel Bucket Wash Mitt to reach under and clean the bit of the arch that sits directly behind the lip roll of the wheel arch (mentioned above). If you don't, even though you can't see it, you'll always know the grime lurks there, gradually eating away at your bodywork! When finished with the Brushes and Mitt on this stage, give the whole area under the arch another thorough rinse down with your Pressure Washer or Hose Nozzle. Move your Snow Foam Cannon or Handheld Foam Sprayer away from the car out of the way.

Stage 4

APC Spray: Now, grab your APC Spray bottle, shake well and generously spray down the under arch area with APC, including the places you can't see, and allow it to soak for two minutes. Take your ValetPro Chemical Resistant Wheel Brush and work the bristles over the whole area, rinsing and lubricating the Brush regularly in your wheel wash bucket as you go.

Utilise your best-matched Wheel Wash Brushes to reach in the nooks and crannies as best you can. When finished with the Brushes on this stage, give under the arch another thorough rinse down with your Pressure Washer or Hose Nozzle. Move your APC Spray away from the car out of the way. Check how dirty your Wheel Wash Bucket water has become; it's probably pretty nasty by now. If so, discard it, rinse the bucket, and refresh it with another half bucket of clean, warm water mixed with one capful of Pure Shampoo.

Stage 5

Fallout Remover Spray: Now we're cooking with gas! This stage is to rid the arches of fallout contaminants that doubtless reside there, and the APC may have missed. Bravely seize your Fallout Remover Spray and put your face mask on as a buffer against the sulphur stench of it. Shake well and generously spritz down the whole under arch area with Fallout Remover and allow to sit for two minutes. Take your ValetPro Chemical Resistant Wheel Brush and work the bristles over the whole area, rinsing and lubricating the brush regularly in your Wheel Wash Bucket as you go. Utilise your best-matched Wheel Wash Brushes to reach in the nooks and crannies as best you can. When finished with the Brushes on this stage, give the whole under arch area another thorough rinse down with your

Pressure Washer or Hose Nozzle. Move your Fallout Remover away from the car out of the way.

Stage 6

Tar & Glue Remover Spray: This is the *penultimate* cleaning stage and will tackle the remaining tar spots that will have flung up from the road and peppered the whole arch area. Take your Tar & Glue Remover Spray, shake well and spray down the whole under arch area and leave it for two minutes.

Take your ValetPro Chemical Resistant Wheel Brush and work the bristles over the whole area, rinsing and lubricating the Brush regularly in your Wheel Wash Bucket as you go. Utilise your best-matched Wheel Wash Brushes to reach into all the nooks and crannies. When finished with the Brushes on this stage, give the whole area under the arch another thorough rinse down with your Pressure Washer or Hose Nozzle. Move your Tar & Glue Remover, Wheel Wash Bucket, Brushes, Pressure Washer, or Hosepipe away from the car out of the way.

Stage 7

IPA Wipe Down: We're at the final cleaning stage - phew! Grab your Car Blower and spend a couple of minutes giving the under arches a good air blast to expel most of the rinse water out of the visible areas and nooks and crannies. Then, to prepare the wheel arches to receive dressing solution/protectant, take your IPA spray diluted 70:30, shake well and generously spritz the entire under arch area, getting the IPA spray into all the tight spots. Next, grab your bag of Downgraded Microfibre Cloths and wipe down the whole arch with a clean Cloth, drying it as much as you can, folding the Cloth as you go. You should find that your seven *relentless* detailing stages thus far have successfully cleaned the whole area beautifully. Your Microfibre Cloth should be coming up satisfyingly clean – most likely, this is the *cleanest* the arches have been since the car left the factory! Move your IPA Spray away from the car out of the way.

Stage 8

Dressing Solution/Protectant: Now for the thoroughly *satisfying* bit where the fruits of your labour start to reveal themselves. Take your CarPro Perl Dressing

Spray, diluted 1:3, shake well, and spritz the Perl over all the plastics and rubbers to be dressed, avoiding getting too much dressing on the suspension and brake components. Then, take a fresh, downgraded Microfibre Cloth and work the Perl dressing in well, folding the Cloth as you work the dressing into the nooks and crannies as best you can. At this point, you may find it helps to get into those narrow areas by taking a flat-ended plastic trim removal tool, wrapping the Cloth around the flat end, and using that to work the dressing into the tight spots.

As well as the *magnificent* restorative, mid-sheen darkening of the finish, the Perl also gives the plastics and rubbers some welcome UV and weathering protection. Due to the amount of abuse the wheel arches routinely receive from day-to-day driving on our lovely UK Autumn and Winter roads, I recommend doing two coats of Perl dressing under the wheel arches. Allow the first coat of dressing to cure for ten minutes, then lay down a second coat of Perl and work it in again with the Microfibre Cloth. You'll notice an even deeper, richer darkening and sheen of the plastics after the second coat – lovely! Move your Perl Dressing Spray and Cloth(s) away from the car out of the way.

Get Snapping

You are DONE! Congratulations! Stand back and admire your newly transformed, *gleaming* arches, now befitting of a *worthy* frame for your badass rims. If you fancy taking some shots to capture the moment of wheel arch detailed perfection, I suggest you have a peek back toward the end of Chapter 1, where I provide hints and tips for achieving cracking driveway photos on your smartphone.

> *Top Tip* Before you replace your front rims on the car (and if you don't need to drive it in the next couple of days), you might want to take a gander at Chapter 6. The next project also requires your wheels off, and as two are already off, and the car is safe and secure, it makes sense to tackle the next project before the wheels go back on.

5 Steps to Safely Jacking up the Car to Replace the Front Wheels

I realise I'm repeating myself here, and I make no apology for doing so. Beware that sloppy use of a Hydraulic Trolley Jack can lead to expensive damage to your pride and joy, or worse, injury to yourself. So, please be *super* focused and careful while you're doing this.

Retrieve All Your Jacking Up Kit

Step 1: Grab your Breaker Bar, Wheel Nut Socket, Locking Wheel Nut Key and Wheel Nuts. Your Wheel Hanger Tool should remain screwed in the front wheel hub hole of the last wheel you removed; check it's still securely screwed in at approximately the 12 o'clock position.

Put your kneeling mats and stool in position but out of the way of your Trolley Jack. Check your wheel chocks are still secure under the two wheels that remain planted on the ground. Check that the handbrake is still firmly on.

Prepare the Front Wheel and Wheel Hub

Step 2: We're home detailers, so we work in a certain way and do things right as far as possible. If you want to make sure your wheel will come free without any wrestling drama next time you need to remove it, here's how:

Grab your APC Spray, Scotch Pad and clean Microfibre Cloth, and give the wheel hub a good spritz with APC. Attack the hub with the Scotch Pad, working it for a minute to clean the hub area. Then wipe it down with the Cloth. Next, hit the hub again with IPA Spray and wipe it down with a Cloth. Repeat the same process on the back of your alloy wheel, where the wheel mating surface attaches to the hub, being careful not to hit the painted finish of the back of the wheel with your scotch pad. You should now have two relatively clean mating surfaces. Then, apply some *Copper Anti-Seize Grease* to your gloved finger and dab it sparingly around the wheel hub (but not in the bolt holes). You don't need a lot; just a few pea-size splodges around the hub mating surface will do the trick; this grease helps prevent seizing and corrosion; it is *not* a lubricant.

Then, give the wheel bolts and threads a quick spray with IPA and wipe them dry with a clean Microfibre Cloth. Next, take your *Aluminium Anti-Seize Paste* and dab a small amount of paste on the nut thread. If your wheel bolts are of the moveable spherical cap variety, take your flat-tipped Plastic Trim Removal Tool, spread a small amount of this paste on the flat tip, and work the paste between the bolt head and moveable spherical cap ring. (The bearing surface of the spherical cap facing the wheel must not be greased). This paste will protect against corrosion and the seizing up of your threaded nuts and bolts. If you're anything like me, you'll find this routine maintenance work both satisfying and therapeutic!

Position the Front Wheel on the Wheel Hub

Step 3: Lift the wheel until you're able to slide it onto the Wheel Hanger Tool from the approximate 12 o'clock position. The wheel hanger will take the wheel's weight, so *you* don't have to. Push the wheel against the hub until you feel it *seat* correctly and the wheel hub holes align with the nut holes on the wheel. Then starting to the right of the wheel hanger, insert the first wheel nut and finger tighten. Next, working clockwise, screw in the other wheel nuts, finger tight in a *star* formation (every other hole – see the diagram below) until you arrive at the Wheel Hanger Tool. Unscrew the Wheel Hanger Tool, replace it with a Wheel Nut, and finger tighten. Two Wheel Nuts will remain, go ahead and screw them both on finger tight.

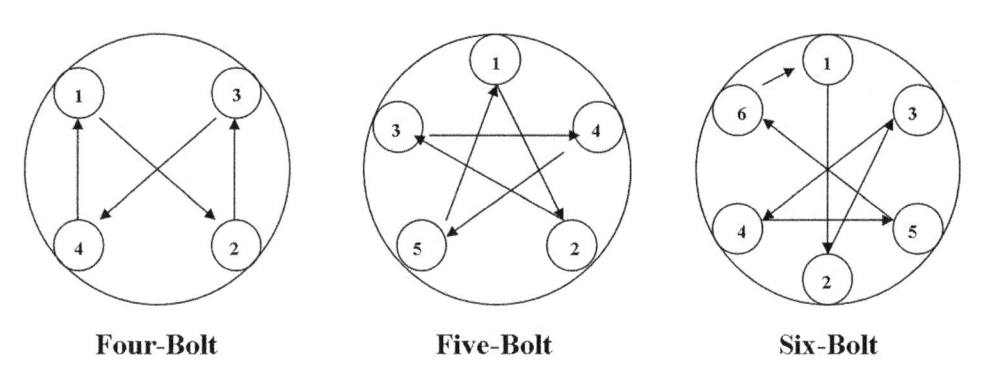

Four-Bolt   **Five-Bolt**   **Six-Bolt**

Once you've got the nuts finger tight, you might find it easier to take the Wheel Nut Socket off the Breaker Bar and continue tightening the nuts just to the *bite* point by holding the Socket head directly in your hand.

My Cayman handbook recommends you then torque all five bolts to 30Nm (22ft-lb) in a star pattern to '*set*' the wheel while the car's still in the air. Set your torque wrench to 30Nm, check it's in the *tighten* position, and torque down all five bolts in the usual *star* formation. 30Nm isn't very tight, so your torque wrench will *click* at you very quickly. You don't finish tightening the wheel bolts until the car is back on the ground.

Lower the Front of the Car Safely

Step 4: Reach under the car and carefully remove the two emergency Axle Stands that were placed loosely underneath the lifting points on each side of the vehicle. Move them well away from the car out of the way. Retrieve your Trolley Jack and carefully line the Jack saddle up underneath one of the now vacant lifting points.

> *Top Tip* Be super aware of how you position the Trolley Jack as there may be occasions when it's not easy to remove it after lowering the car. Or, if the Trolley Jack positioning is careless, the jack handle could damage part of the undercarriage as the vehicle is lowered – ouch!

When you're happy that the Trolley Jack saddle is correctly aligned, carefully and slowly start to lift the car clear of your primary Axle Stand. Keep a keen eye on the jack saddle's contact with the lifting point as you lift. If the saddle starts to slip, *STOP* immediately, lower the car and reposition the jack saddle. You don't need to raise the car very high, just enough to be able to slide the Axle Stand out and well clear of the vehicle. When the Axle Stand is well clear of the vehicle, *slowly* lower the car back onto the ground, then move your Trolley Jack out of harm's way. Chock both sides of the front wheel that's now back on the ground and repeat the process on the other side of the car.

Complete Wheel Bolt Tightening

Step 5: To complete tightening the front wheel bolts, for most Porsche models, set your torque wrench to **130Nm** (96ft-lb) (check your vehicle handbook for your correct torque setting). Take your Locking Wheel Nut Key and Torque Wrench. Check that your Torque Wrench is set to *tighten*, and starting from approximately the 12 o'clock position, torque down all five bolts in the usual *star* formation. Be careful not to hit the Torque Wrench handle against the side of the car. When all five bolts are tightened to 130Nm, *or your correct torque setting,* finish by applying a small squirt of WD-40 to the bolt's heads to help protect against corrosion.

> *Top Tip* This safety tip comes courtesy of Golding Barn Garage, my trusted local Porsche indie. Set a diary note to check your wheel bolt torque settings

after you've driven 50 miles. A 5-minute job for peace of mind is 5 minutes very well spent!

Let's Get the Rear Arches Detailed

Raising the Rear of the Car: When you've replaced the front wheels on the car, double-check you've chocked the front and back of each front wheel, once again using two chocks per wheel.

Then, move to the rear of the car and jack up the rear end the same way you did the front. If you have access to a Long Reach Trolley Jack, lifting the rear is a bit easier than the front.

I like to jack up the rear using the rear suspension mounting point located at the mid-way point of the undercarriage, between the two rear wheels. This spot attaches the suspension to the rear chassis and is easily strong enough for lifting purposes.

Jacking the car up from this position will usually raise both rear wheels off the ground together. This scenario is ideal as it enables you to set each Axle Stand to the same height and put them carefully and securely in place on both sides of the car under the factory standard jacking points at the same time. *Do not attempt to lift the vehicle using any part of the engine or gearbox.* Carefully and slowly lower the car onto the Axle Stands and remove the trolley jack well away from the car out of the way.

Top Tip At this point, I like to put in the other safety precaution. See the Top Tip on Page 70 regarding the emergency third and fourth Axle Stands.

Check the Car's Stability

Follow the same process described in Step 4 on Page 70.

Rear Wheels Removal

Follow the same process described in Step 5 on Page 71.

8 Stages to Wicked Wheel-Arch Detailing and Protection (rear)

Follow the same 8-stage process described in steps 1-8 on Pages 71-75.

Get Snapping

This is optional; see Page 75.

5 Steps to Safely Jacking Up the Car to Replace the Rear Wheels

Follow the same five steps to replace the rear wheels described in steps 1-5 on Pages 75-78.

Clear Up

I know, I know, doing the clear up is boring; you might be tired and can't be bothered. As I keep banging on, clear up is a necessary evil and will ensure all your precious detailing and maintenance kit stays in tip-top condition, ready to go next time you need it. Start by returning all your Liquids and Spray Bottles to storage, but do check all lids and spray heads are secure and keep an eye on any dilutables that need topping up. There's nothing worse than going for your detailing spray, only to find there's just a dribble left! If you used the Handheld Foam Sprayer, rinse it out with clean water, semi-pressurise it, and spray out some clean water to clear the feed pipe and nozzle. Give it a quick dry and store it away. Grab your Car Blower, Caddy, Mats, Stool, Detail Guardz, Wheel Chocks and give them a quick wipe down to dry if necessary and store them away. Give your Trolley Jack and Axle Stands a quick wipe down with WD-40 and store them away.

Return your Locking Wheel Nut key to its place of safekeeping. Release the tension on your Torque Wrench and store it in its box. Never store the torque wrench under load as this will reduce the tool's life.

Discard all the bucket water and give your Wheel Bucket and Grit Guard a rinsing blast with the Pressure Washer or Hose Nozzle but hold back the Wheel Bucket for now as you'll need it a bit later. Next, gather your used Cloths and Mitt and squeeze them out if necessary. The Cloths are likely to be very mucky from the wheel arch duty, so chuck them in your Wheel Bucket, half fill it with clean, warm water, add a small squirt of washing-up liquid and give them a short pre-wash with your gloved hand. Then squeeze them out, discard the dirty wash water, give the Wheel Bucket a quick rinse, and pop the Cloths and Mitt in the washing machine at 40°C (104°F). Add the Non-Bio Microfibre Detergent and stick on a 30-minute cycle with a gentle spin. When you take them out of the machine, give everything a good shake to re-set the fibres, and hang everything on a clothes dryer to dry naturally, using a peg for the Wheel Mitt.

You will be left with just your Pressure Washer, Wheel Brushes and Wheel Bucket. Dump all your Brushes in the Wheel Bucket and give them a quick rinsing blast with the Pressure Washer or Hose Nozzle. Discard that rinse water in the water butt if you have one. Add a squirt of washing-up liquid to the bucket and half-fill it with hot tap water, swirling it around to mix. Washing-up liquid is an excellent degreaser and is perfect for brush cleaning in this way. Spend a couple of minutes with your gloved hands in the soapy water, massaging the Brushes clean. Dump the dirty soapy water and give the Brushes a final rinsing blast in the Bucket with the Pressure Washer or Hose Nozzle, then dispose of that rinse water in the water butt.

I then give all the Brushes a good wrist flick to fling off some of the excess rinse water, and I lay them on an old dry towel under the radiator to dry, or in the sun if you're lucky. Give your Wheel Bucket a quick rinse, wipe it dry and store it away with the clean Grit Guard. Give your Pressure Washer and accessories, including the power cable, a wipe down to dry and store them away. Next day when the Cloths, Mitt and Brushes are fully dry, store them away, keeping your Wheel Brushes in your Wheel Wash Bucket ready for next use.

Reflection

Congratulations on completing your wheels off-wheel arch detail and protection project! Perhaps, only a true *Driveway Warrior* will understand just how immensely satisfying it feels. If you did capture the moment of arch detailing perfection on your phone, I bet you keep returning to those pics to remind yourself what an utterly fabulous job *you* did, and quite rightly so! Time to chill and enjoy those beers on ice while you reflect on your awe-inspiring arches and the £240 ($321) you have saved on what the pro detailer would have charged you for the job you've so brilliantly done yourself. You are *awesome*!

Once rest and relaxation are underway, I wonder if your thoughts may turn to giving your *stunning* rims and brake calipers the TLC and long-term protection they deserve? If two of your wheels are already off the car, why not *seize* the opportunity to give them some serious loving with a *Spectacular* Ceramic Coating Project.

"How?" I hear you cry.

Well, that would be *Driveway Detailing Warrior style,* of course. Read on in Chapter 6, and I'll explain precisely how ...

CHAPTER 6

Wonderful Wheels off - Alloys, Brake Calipers & Exhaust Tip Detailing & Ceramic Coating

They are totally flipping *awesome*, right? I'm talking about your wheels, your alloys, your rims. Of course, they are, be they 15in, 18in, or 22in, it's all the same; they all *rock*! However fancy your car is, whatever it cost, be it £5K or £100K, your alloy wheels are one of the focal points of your ride, and they will always grab the attention of the beholders. But how best to keep them looking *fabulous* while also protecting them from the elements? If you want to give your expensive alloys the love and care they deserve, get them ceramic coated, and while you're at it, why not give the same loving to your brake calipers and exhaust tips! Luckily for us, excellent ceramic coatings are now affordable and available to all.

This is where the diminutive but *formidable* bottles of *Gtechniq's C5 Wheel Armour* come to your service. C5 uses a scientific formula to chemically bond the ceramic coating to your rims, making them repellent to brake dust, muck and contamination. This makes washing them *oh* so much easier, and with up to two years of durability, a blast with the Pressure Washer or Hose Nozzle is usually sufficient to keep your alloys *gleaming*. In my experience, the non-ceramic variety of traditional wheel wax sealants just can't match the durability of C5, and that's why I go for C5.

Avoid the Sun

Like a red-haired granny in mid-August Benidorm, home detailers need to avoid the sun at all costs. May I refer you to Chapter 1, Page 2, *'Sunshine – the Safe Wash Enemy.'* Never start or finish your Detailing Processes in direct sunlight if you want to avoid issues with your detailing liquids drying out too quickly on your car and *screwing* up the quality of your finish. Pick a couple of days when the forecast is dry and mild, preferably overcast when your paintwork is cool to the touch. We want to give ourselves every opportunity to achieve the perfect results, right?

Wheels off Rims, Calipers & Exhaust Tip Ceramic Coating Kit List

Detailing Hardware Required:

- Carry Caddy
- Nitrile Gloves
- Eye Protection
- Kneeling Mats
- Folding Stool
- Pressure Washer Kit & Extension Lead *or* Hosepipe & Nozzle
- Wheel Wash Bucket & Wheel Kit Brushes
- Car Blower
- Detail Guardz
- Handheld Foam Sprayer (diluted 1:4 APC to Water if not using a Pressure Washer)
- Microfibre Cloths (a bag full of downgraded cloths)
- Trim Removal Tools
- C5 Mini Applicator Pads
- Extra Fine Steel Wire Wool (for chrome exhaust tips only)
- Clay Mitt
- Polish Applicator Pad
- Scotch Pad

Top Tip C5 Wheel Armour comes with a few mini applicator pads, but not enough to do your wheels, calipers and exhaust tip(s). You can get more of these mini pads pretty cheaply from eBay. The C5 comes in 15ml and 30ml bottles; you'll need the 30ml to do the *holy trinity* of wheels, calipers and tips!

Detailing Liquids Needed:

- Rust Repellent (diluted 1:3)
- Pure Shampoo
- Snow Foam (diluted 1:4)
- All-Purpose Cleaner (APC) Spray (diluted 1:4)
- Isopropyl Alcohol (IPA) Spray (diluted 70:30)
- WD-40

- Fallout Remover Spray
- Tar & Glue Remover Spray
- C5 Wheel Armour
- CarPro Perl Dressing (diluted 1:2)
- Peek Metal Polish (for chrome exhaust tips only)
- Autoglym Super Resin Polish
- Microfibre Detergent

To do this job properly and get the maximum from the C5, you'll need to remove your wheels, either one or two at a time, depending on what kit you have. This is no problem, though the procedure for lifting a 1430 kg (3153 lbs) vehicle can be a little daunting for the Driveway Warrior, particularly the first time. However, with the right tools and observing the basic safety steps covered in the following pages, safe wheel removal is a basic procedure you CAN tackle. Since the drama of my initial efforts years ago using an inadequate jack (I expand on this later in the chapter), I have done so dozens of times on my driveway without incident, and you can do it as well.

Wheel Removal & Replacement Tools:

- Two Sets of Axle Stands (Jack Stands), i.e., four in total *or* a Rennstand Safe Jack
- Low Profile Hydraulic Trolley Jack
- Wheel Chocks x 4
- Breaker Bar & Wheel Nut Socket (most Porsche wheel nuts are 19mm, check the handbook for your correct size)
- Torque Wrench
- Wheel Hanger Tool
- Locking Wheel Nut Key
- Container for Wheel Nuts
- Copper Grease
- Aluminium Paste

Top Tips Use the plastic-sheathed variety of wheel nut socket head when removing and replacing your wheels. The plastic coating over the socket head will help prevent damage to your precious alloys. Wrap some insulation tape around the head of your locking wheel nut key to help prevent the key from scuffing against the inside of your wheels while working with the locking key.

Splitting Up the Work

Realistically, there's a full day's work involved in detailing and ceramic coating all four wheels, plus brake calipers and exhaust tip(s). I prefer to split the work up over two days by doing:

Day 1: Do the two *front wheels* plus front calipers, allowing those wheels and calipers to cure for 12 hours.

Day 2: Replace the two front wheels on the car and then do the two *rear wheels*, the back two calipers plus exhaust tips (access to the exhaust tips is much easier when the car's rear is in the air).

Preparation

If you've already done the Chapter 5 project and still have a couple of wheels off the car, you're laughing! You can skip steps 1-5 and go straight to stage 1 on Page 91. Remove your rings and watch, and make sure you aren't wearing anything zippy or with buttons or poppers that could scratch your car while you're working up close to it. As always, remember to get those beers in the fridge, ready for when you finish, and grab a bottle of water or Thermocafé mug of coffee to sustain you.

Position your car on a *level, flat, stable* surface (please no sloping driveways for this project), preferably with room to safely work around it.

Place your Detail Guardz under the outer edges of the tyres of the wheels not being removed. Get your Ceramic Coating detailing kit and wheel removal/replacement tools assembled. Load up your Carry Caddy with your Spray Potions, Microfibre Cloths, Wheel Hanger Tool, Breaker Bar, Wheel Nut Socket, Locking Wheel Nut Key, etc., and have your Caddy handy. Put your Kneeling Mats and Stool in position but out of the way of your Trolley Jack.

A Word on Low Profile Hydraulic Trolley Jacks

I use the *Halfords Advanced 2 Tonne Low Profile Hydraulic Trolley Jack* to lift my Cayman; it has a solid, robust feel, which meets all my needs. See my summary in Chapter 5 on Page 66.

I do not recommend the smaller style of mini trolley jack for this task, and I regard these mini jacks as suitable for *emergency* use only. The reason being the jack *saddle* tends to be much smaller, and therefore its contact area under the lifting point is somewhat limited. Years ago, I attempted to lift my car using one of these mini trolley jacks, and I experienced a *scary* moment when the jack saddle partially slipped off the jacking point. Disaster was narrowly averted, but I learned an important lesson about always using the correct tools.

Top Tip If you're unfamiliar with your Trolley Jack, please see the Top Tip in Chapter 5 on Page 67.

Day 1

5 Steps to Safely Jacking Up Your Car & Removing the Front Wheels

Make no mistake, incorrect use of a Hydraulic Trolley Jack can lead to expensive damage to your pride and joy and/or worse, injury to yourself. The alternative is to pay a pro detailer around £291 ($390) to do this project for you. But if you have the kit, a safe area to work, and the confidence to have a go, let's *DO* this.

Before You Raise the Car

Step 1: I repeat, make sure the car is on a flat, level, stable surface; once again, please *no* sloping driveways for this project. Remove any excess weight from the vehicle, e.g., spare wheel or children! Locate the Locking Wheel Nut Key, remove the ignition key, firmly apply the handbrake. If you place the vehicle in park (for an automatic gearbox/transmission), it will only lock the two rear wheels. Check that your Axle Stands are all in good condition; if anything looks badly corroded, bent, or damaged, please *bin* it immediately; it simply *isn't* worth the risk of using any defective kit!

Chock the Wheels & Loosen the Nuts

Step 2: With your Gloves on, carefully chock the front and back, both sides of each rear wheel; this is an extra precaution against the car rolling. So, if you're raising the front of the vehicle to remove the front two wheels, you need to chock both sides of the same two rear wheels (two Chocks per wheel). The reason for doing this is that most surfaces are not completely level - as always, safety first, friend!

While the wheels are still on the ground, use the Breaker Bar and Wheel Nut Socket to 'crack' open (loosen one turn) all five front wheel nuts, working anti-clockwise on each front wheel. If you don't do this now, your wheels will spin, and you'll have a headache trying to loosen the wheel nuts. Don't forget you'll need the Locking Wheel Nut Key for your locking wheel nuts. Most cars have one locking wheel nut on each wheel.

Warning *Never, under any circumstances, work on a raised car just* using *the* trolley jack alone, *without axle stands being in place to support the car's weight. Not even if it's just for a minute. Also, be aware that the handbrake will no longer work when you raise the car's rear as it only works on the two rear wheels.*

Raising the Front of the Car

Step 3: This is the longest step as there's a fair bit of safety stuff to take care of to ensure Driveway Warriors will battle on another day! Consult your owner's handbook to identify the car's four factory standard jacking points. These are the strengthened sections on the chassis where it's safe to put either your jack saddle or Axle Stands. However, the obstacle here is that there isn't enough room to fit *both* your Trolley Jack lifting saddle *and* your Axle Stand together under the same jacking point. This problem is eliminated if you're lucky enough to own a *Rennstand Safe Jack*: a great bit of kit that enables both safe lifting *and* supporting together at the exact correct spot. Unfortunately, at the time of writing, my understanding is that the *Rennstand* is still only available in the USA. You can order one direct from Safe Jack in the USA, though you'll obviously have to fork out for shipping wherever you are. I paid about £70 ($93) to ship my Rennstand to the UK if my memory serves me correctly.

So, how to get around this challenge without a Rennstand? I prefer to lift the front end of the car first, and I only remove the front two wheels before *replacing* them on the vehicle and *lowering the front* of the vehicle before going on to raise the rear of the car. The car is far more stable, with two wheels remaining planted on the ground. The Cayman has a reinforced area of the front chassis, just behind the front factory standard jacking point, which is the ideal point to jack the car up. If you place your trolley jack under this section to lift, you'll have enough room to fit your Axle Stands under the factory standard jacking point at the same time. Don't place the Axle Stands under any part of the engine or gearbox, as it may make the car unstable.

Take your Trolley Jack and line up the jack saddle underneath the reinforced area of the front chassis, just behind the factory standard jacking point. Have your Axle Stand ready beside you, and raise the jack saddle slowly, checking it's lined up perfectly under the reinforced area. I like to place a folded Microfibre Cloth on the jack saddle to prevent metal on metal contact from scraping against the undercarriage. Lift the car slowly until the wheel clears the ground by approximately 2cm (1in).

It's normal for the car to tilt slightly while the wheels on the opposite side remain on the ground. You only need to lift the vehicle enough to enable you to slide the Axle Stand set to its minimum height securely under the factory standard jacking point. Carefully line it up under the factory standard jacking point. Once the Axle Stand is in position, carefully and slowly lower the car onto the Axle Stand and remove the Trolley Jack away from the car out of your way.

> *Top Tip* At this point, I like to put in another safety precaution. See the Top Tip in Chapter 5 on Page 70 regarding the emergency third and fourth Axle Stands.

Repeat the same process on the opposite side of the car, so you have both front wheels off the ground and supported with Axle Stands. Ensure both Axle Stands are set to the same height so that the car sits level and secure.

Warning *Never put your fingers between the car's jacking point and the axle stand during the lifting process.*

Note: *When you start to lift the car, it's important to be aware that depending on where you place your trolley jack saddle, it's possible both wheels on the same side of the car could rise off the ground, or both front wheels may raise off the floor at the same time. However, only the wheel nearest the trolley jack will rise if you use the above-recommended lifting point.*

Check the Car's Stability

Step 4: We need to do one final safety check before removing the front wheels. Now, gird your loins, gather your faith and courage, and standing to the side at the raised end, give the car a firm shove or two to check if it's super stable on the Axle Stands.

If the car doesn't budge, you're good to go. But if the car moves at all, you do *not* have it correctly supported. It's far better that the vehicle falls off the Axle Stands while you're shoving it from the side than it falls on any part of you while you're working on it!

Front Wheels Removal

Step 5: Finally, it's safe to remove the front wheels – phew! So, you've already loosened the wheel nuts in Step 2. Grab your Breaker Bar, Locking Wheel Nut Key, Wheel Nut Socket and Wheel Nut Container, and further loosen all five Wheel Nuts on one wheel. Once you've got the nuts loosened to a certain point, you might find it easier to take the Wheel Nut Socket off the Breaker Bar and finish removing the nuts by holding the Socket directly in your hand.

> ***Top Tip*** First, remove the wheel nut closest to 12 o'clock, then screw your wheel hanger tool securely through the vacant wheel nut opening into the wheel nut hole on the wheel hub. Finish removing the other four wheel nuts and put all five nuts safely in your container. Grasp the wheel firmly with two hands and give it a hefty wiggle. When the wheel comes free of the hub, it will drop *blissfully* onto your Wheel Hanger and not crash sickeningly down onto your painted brake caliper, causing a horrendous chip in the process!

Slide the wheel off the Wheel Hanger and put it aside somewhere safe to work on later. Move your Lifting Gear and Wheel Removal Tools away from the car somewhere safe, out of the way.

10 Stages to Front Rims, Calipers & Exhaust Tip Detailing & Ceramic Coating

Grab your Caddy loaded with your Detailing Products and Wheel Brushes, and let's get into some actual detailing! Take your Wheel Bucket with Grit Guard at the bottom, add two capfuls of Pure Bodywork Shampoo and fill with warm water; add two capfuls of APC for extra foaming and cleaning power. Swirl the water around to mix in the Shampoo and APC.

Front Wheels & Calipers Decontamination

Stage 1

Foam Pre-Soak: Having removed the two front wheels, with your Gloves on, take your Rust Repellent spray, shake well and give each brake disc (rotor) a good spray down. Allow it to cure while you work on the wheels and calipers. Then move your Rust Repellent away from the car out of the way.

Prop up the two wheels you're working on somewhere so that they're reasonably stable. If the wheels have never been detailed or sealed before, the backs will likely be grim with baked-on brake dust and muck. If using a Pressure Washer, generously coat each wheel and tyre back and front with Snow Foam and allow to soak for five minutes. If not using a Pressure Washer, take your Handheld Foam Sprayer, shake well and liberally foam up each wheel and tyre with APC Foam and leave for five minutes. While the Foam is dwelling on the wheels, foam up the calipers in the same way. After five minutes, give the wheels and calipers a thorough rinse down with your Pressure Washer or Hose Nozzle.

Stage 2

Brush Attack with Foam: Next, give the wheels and calipers a second foam bath with Snow Foam or Handheld APC Foam and allow them to soak for two minutes. At this point, you're likely to be working the brushes at eye level, so safety first; please put on your eye protection.

While you're working on the front calipers, if you have a front-wheel drive vehicle, you may be able to carefully turn the brake assembly toward you for ease of access.

Take your Wheel Wash Bucket & Kit Brushes and, using your Kneeling Mats and Stool if you wish, use your brush selection to get scrubbing! Start with the calipers as they likely hold less dirt, then turn your attention to the wheels. Use the Meguiar's Supreme Brush to start with – this one absorbs a *fearsome* amount of water. Rinse it regularly in the Wheel Wash Bucket, and then work your way through your Wheel Brush Kit. Keep them all super-well-lubricated with wash water (a dry brush is a scratchy one).

Your ValetPro Chemical Resistant Wheel Brush is likely to be the most effective at attacking layered dirt for the wheel backs. Give this Brush a turbo boost by spraying more APC on the bristles before you go at the back of the wheels. Give the inner and outer tyre walls a good going over with your Tyre Brush until any brown residue diminishes. When finished with the Brushes on this stage, give the calipers and wheels a thorough rinse with your Pressure Washer or Hose Nozzle. Move your Snow Foam Cannon or Handheld Foam Sprayer away from the car out of the way.

Stage 3

Brush Barrage with APC Spray: Now, grab your APC Spray bottle, shake well and generously spray down the calipers and wheels front and back and allow to soak for two minutes.

Utilise your best-matched Wheel Wash Brushes to reach in the nooks and crannies as best you can. Take your ValetPro Chemical Resistant Wheel Brush and work it all over the calipers before going at the wheels again, rinsing and lubricating the Brush regularly in your Wheel Wash Bucket as you go. When finished with the Brushes on this stage, give the calipers and wheels another thorough rinse with your Pressure Washer or Hose Nozzle. Move your APC Spray away from the car out of the way.

At this point, check how dirty your wheel wash bucket water has become. It's probably pretty nasty by now, so discard it, rinse the Bucket and Grit Guard, and refresh it with another bucket of clean, warm water mixed with two capfuls of Pure Shampoo.

Stage 4

Brush Frenzy with Fallout Remover: Now, we can check what fallout material remains on the wheels and calipers. Put your face mask on as a buffer against the Fallout Remover sulphur stench and bravely seize your Fallout Remover Spray. Shake well and generously spritz down the calipers and wheels, front and back, with Fallout Remover and allow to soak for two minutes while you observe the satisfying purple chemical reaction. Utilise your best-matched Wheel Wash Brushes to reach in the nooks and crannies as best you can. Take your ValetPro Chemical Resistant Wheel Brush and work it over the calipers, followed by the wheels, rinsing and lubricating the brush regularly in your wheel wash bucket as you go. When you have finished with the brushes on this stage, give the calipers and wheels another thorough rinse down with your Pressure Washer or Hose Nozzle. Move the Fallout Remover away from the car out of the way.

Stage 5

Brush Assault with Tar & Glue Remover: This is potentially the penultimate cleaning stage and will tackle any remaining tar spots that have flung up from the road and peppered the wheels and calipers. Take your Tar & Glue Remover Spray, shake well and generously spray down the calipers and wheels front and back and allow to soak for two minutes. Then, take your ValetPro Chemical Resistant Wheel Brush and work it over the calipers, followed by the wheels, rinsing and lubricating the brush regularly in your wheel wash bucket as you go. Utilise your best-matched wheel wash brushes to reach in the nooks and crannies as best you can. When finished with the brushes on this stage, give the calipers and wheels another thorough rinse down with your Pressure Washer or Hose Nozzle. Move your Tar & Glue Remover away from the car out of the way.

At this point, we need to dry the calipers and wheels to assess the level of contaminants remaining. Grab your Car Blower and give them all a good air blast to expel the bulk of the rinse water out of the nooks and crannies, then dry them off using clean, dry Microfibre Cloths. Taking your Gloves off, run your fingertips lightly over the calipers and wheel's surface. Do they still feel a bit rough, almost like fine sandpaper? If they still feel a bit rough, contaminants still linger on the surface, so move on to Stage 6, *'Clay Mitt.'* But if they feel pretty smooth, skip Stage 6 and move on to Stage 7, *'To Polish or Not,'* or if you don't need to polish, cut straight to Stage 8, *'Final Prep.'*

Stage 6

Clay Mitt: This stage is if contaminants linger, making the calipers or wheels still feel a little rough or gritty. With your Gloves back on, snag a clean Bucket and Grit

Guard, add a capful of Pure Shampoo, and half-fill it with warm water. Swirl the water to mix in the shampoo – this will be your Clay Mitt lubricant. Grab your Clay Mitt and dunk it into the soapy water to lubricate it thoroughly. Then, gently work the Clay Mitt over the calipers and wheels using *light pressure* only, keeping the Clay Mitt super-well-lubricated with soapy water (a dry Clay Mitt may dull the paint finish). The Mitt will help to safely remove bonded contaminants, ingrained dirt, and tar spots. If your brake calipers have logo stickers on them, be super careful the clay mitt doesn't damage the sticker.

Next, take your Pressure Washer or Hose Nozzle and rinse down the calipers and wheels thoroughly. Finish this step by blasting the excess rinse water off with your Car Blower, then take some clean, dry Microfibre Cloths and dry off the calipers and wheels, ready for the last cleaning stage. With your Gloves off, run your fingertips over all the surfaces to check they're now lovely and smooth. Finally, move your Pressure Washer or Hosepipe, Wash Bucket, Brushes, Clay Mitt and Car Blower away from the car out of your way.

Stage 7

To Polish or Not: Before you go any further, step back and examine the finish of your brake calipers and the front faces of your alloys. Are you happy with the finish? Does it gleam vibrantly, or is it *just a* little bit flat? If you're satisfied with the finish, skip the rest of Stage 7 and go to Stage 8. If it's a little bit too *dull* for your liking, here's how to sort it.

Gloves back on, grab your *Super Resin Polish* (its gentle abrasives won't hurt your calipers or wheels), Polish Applicator Pad and a clean, dry Microfibre Cloth. Shake the Polish well and apply two pea-sized dots to the Applicator Pad.

Pick your starting point and work a thin layer of polish onto the wheels with moderate pressure using a circular motion. Do this in sections in an overlapping crosshatch pattern for one minute. Be careful to apply even pressure over the Pad to avoid pressing down with your fingertips in a concentrated area.

Allow it to dry for a couple of minutes before buffing off your worked section with the Microfibre Cloth, folding the Cloth frequently as you go. Go over the whole wheel in this way, applying more blobs of polish to your applicator pad as required. Remember, less is more with this polish; it will start to 'dust up' and become more difficult to buff off if you apply too much.

Try not to get polish residue between the wheel rim and the tyre. If you do, straightaway take your clean, dry Exterior Detailing Brush, work it into the rim/tyre gap, and it should remove the polish residue, provided it hasn't gone hard. If it has gone hard, grab your flat-ended trim removal tool, fold a clean Microfibre Cloth over the flat end and ease out the irksome residue. Apply a pea-size blob of Polish to the Applicator Pad to do your calipers and buff them up to shine. If you get Polish in any of the caliper nooks and crannies, no problem; just take your clean, dry Exterior Detailing Brush and work the polish residue out with the bristles. Move your Super Resin Polish, Applicator Pad, used Polish Cloths and Brush away from the car out of your way.

Stage 8

Final Prep - IPA Wipe Down: You might think this stage is overkill, but I suppose that's the nature of detailing, and this is the crucial final step before you break out your C5 ceramic coating. Take your IPA Spray bottle, diluted from the 99.9% stuff to a 70:30 mix (IPA to De-ionised Water), giving you the ideal 70% proof IPA panel wipe. Shake well, spritz the calipers with IPA, and then wipe them dry with a clean, dry Microfibre Cloth. This will give your ceramic coating the best possible chance to bond to the paintwork and do its magic of achieving up to 24 months of protection. The last IPA wipe down will remove any chemicals or oils lingering on the surface from the cleaning and polishing stages or from your hands - leaving the surface as clean and bare as possible.

Grab a couple of different coloured clean, dry Microfibre Cloths. Evenly spritz the IPA over the front and back of the wheels. Then, spread the IPA over the wheel with one Cloth and wipe it dry with the other, folding your drying Cloth as you go.

The IPA leaves no residue and will flash away quickly with the help of your drying Cloth. While you're at it, spray down the inner and outer tyre walls with IPA and wipe them dry with a downgraded Microfibre Cloth. This will help the Perl tyre dressing bond to the tyre walls giving the dressing excellent longevity. Move your IPA Spray and used Microfibre Cloths away from the car out of your way when finished.

Stage 9

Dress the Tyre Walls: On to the tyre dressing: take your CarPro Perl spray diluted 1:2, shake well and evenly spray a coating right around the tyre wall. Then using an old Microfibre Cloth, work the dressing into the tyre wall. Repeat for the inner and outer tyres walls. For a month or more of durability, allow the first coat of dressing to cure for ten minutes before applying the second coat of Perl and work it in with the Cloth. It will give you a lovely, durable mid satin sheen that will last several weeks, and I've found it will not sling off all over your paintwork as soon as you drive the car.

Stage 10

5-Step C5 Wheel Armour Application: If at all possible, at this point, you should try and move your front wheels inside or undercover. This will avoid the external environment affecting your wheel application, e.g., potential rain or dust, both while you're applying the C5 and while it cures for 12 hours. If you *can* move the wheels inside, just make sure the room is well ventilated by opening windows or doors. Ideally, you don't want to be down on your knees applying the C5, so try to get the wheels up on a workbench where you can work in comfort and give them your best effort!

So, all that done, let's get cracking on the C5 application steps:

1) Load up your Caddy with your C5, Mini Applicator Pads, and clean, soft buffing Microfibre Cloths. Put your Gloves and Eye Protection on; you don't want this stuff on your hands or in your eyes. Check that the C5 lid is on tight and shake well; put a few drops of C5 across an area the size of a £1 coin on the supplied Applicator Pad.

> *Top Tip* Replace the lid on the bottle and set the bottle aside (keeping the bottle held in your hand will accelerate the precious C5, curing wastefully in the bottle).

2) Working first on the two front calipers, carefully spread the C5 over every bit of the calipers, including the backs and underneath you can't see. Then take your clean, dry Microfibre Cloth and gently flatten the C5 over the calipers. Don't buff

it hard; you're not trying to buff it off; just gently run your Cloth over the calipers to flatten any high spots, folding the Cloth as you go to help prevent the C5 curing on the Cloth.

3) Next, move inside or undercover to do the wheels. Working first on the back barrels of the wheels, take a fresh Mini Applicator Pad, apply more drops of C5, and methodically spread it over every bit of one-quarter of the back barrel. Then take your Microfibre Cloth and gently flatten the C5 over your working area. Don't buff it hard; remember you're not trying to buff it off. Just gently run your Cloth over the area to flatten any high spots, folding the Cloth as you go.

4) It's difficult to see where you've done as the C5 is very translucent, and that's why you need to apply it *methodically* to avoid waste by doing the same area twice

or missing any spots. Apply more C5 on the Applicator Pad and go again over the second *quarter* of the barrels before gently flattening with the Microfibre Cloth.

Continue until you've done the whole rear barrel before turning the wheels over, changing to a fresh Mini Applicator Pad, and repeating the same steps on the front faces. I found that just over *half* a 30ml bottle did my trio of four wheels, four calipers and exhaust tip. That said, I was careful not to overapply it wastefully.

5) Allow to cure for 12 hours, indoors or undercover, before replacing the front wheels on the car on day 2.

Day 2: Replace the Front Wheels on the Car

5 Steps to Safely Jacking Up the Car to Replace the Front Wheels

At the start of day 2, your *fabulous* newly ceramic coated front wheels will have cured for 12 hours and be ready to go back on the car.

I realise I'm repeating myself here, and I make no apology for doing so. Please beware that sloppy use of a Hydraulic Trolley Jack can lead to expensive damage to your pride and joy and/or worse, injury to yourself. So, please be *super* focused and careful while you're doing this.

Retrieve Your Jacking Up Kit

Step 1: Grab your Breaker Bar, Wheel Nut Socket, Locking Wheel Nut Key and Wheel Nuts. The Wheel Hanger Tool should still be screwed in the wheel hub hole of the last wheel you removed. Check it's still securely screwed in at approximately the 12 o'clock position.

Put your Kneeling Mats and Stool in place but out of the way of your Trolley Jack. Check that your Wheel Chocks are still secure under the two wheels that remain planted on the ground. Ensure the handbrake is still firmly on.

Prepare the Front Wheel and Wheel Hub

Step 2: You only need this section if you skipped Chapter 5. We're home detailers, right, so we work in a certain way and try to do things correctly. If you want to make sure your wheels will come free without any drama next time you need to remove them, here's how:

Grab your APC Spray and Scotch Pad, give the wheel hub a good spray over with APC and attack the hub with your Scotch Pad, working the Pad for a minute to clean the hub. Then wipe it down with a clean Microfibre Cloth. Next, take your IPA Spray, hit the hub again with IPA, and wipe it dry with a clean Microfibre Cloth. Repeat the same process on the back of the wheel, where the wheel mating surface attaches to the hub. Be careful not to hit the painted finish of the back of the wheel with your Scotch Pad. You should be left with two relatively clean mating surfaces. Then, apply some *Copper* Anti-Seize Grease to your gloved finger and dab it sparingly around the wheel hub (but not in the bolt holes). As its name suggests, this grease helps prevent seizing and corrosion; it is not a lubricant. You don't need a lot, just enough for a few pea-size splodges around the hub mating surface.

Then, give the wheel bolts and threads a quick spray with IPA and wipe them dry with a clean Microfibre Cloth. Next, take your *Aluminium* Anti-Seize Paste and dab a small amount of paste onto the nut thread. If your wheel bolts are of the moveable spherical cap variety, take your flat-tipped Plastic Trim Removal Tool, spread a small amount of this paste on the flat tip, and work the paste between the bolt head and moveable spherical cap ring. (The bearing surface of the spherical cap facing the wheel must <u>not</u> be greased). This paste will protect against the corrosion and seizing of your threaded nuts and bolts. If you're anything like me, you'll find this routine maintenance work both satisfying and therapeutic!

Position the Front Wheel Back on the Hub

Step 3: Lift the wheel until you're able to slide it onto the wheel hanger tool from approximately the 12 o'clock position. The wheel hanger will take the wheel's weight, so *you* don't have to. Push the wheel against the hub until you feel it *seat* correctly, and the wheel hub holes align with the nut holes on the wheel. Then, starting to the right of the wheel hanger, insert the first Wheel Nut and finger tighten.

Next, working clockwise, screw in the other Wheel Nuts, finger tight in a *star* formation (every other hole) until you arrive at the Wheel Hanger Tool. Unscrew the Wheel Hanger Tool, replace it with a Wheel Nut, and finger tighten. You should be left with two Wheel Nuts; go ahead and screw them both on finger tight. Once you've got the Nuts finger tight, you might find it easier to take the wheel nut socket off the breaker bar and continue tightening the nuts just to the *bite* point by holding the socket head directly in your hand.

My Cayman handbook recommends you torque all five bolts to 30Nm (22ft-lb) in a star pattern to *'set'* the wheel while the car is still in the air. Set your torque wrench to 30Nm, check it's in the *tighten* position, and torque down all five bolts in a *star* formation. 30Nm isn't very tight, so your torque wrench will *click* at you very quickly. See the Wheel Nut tightening order diagram on Page 77.

You don't finish tightening the wheel bolts until the car is back on the ground.

Lower the Front Safely

Step 4: Reach under the car and carefully remove the emergency third and fourth Axle Stands that were placed loosely underneath the lifting points on each side of the vehicle. Move them well away from the car out of the way. Retrieve your Trolley Jack and carefully line the jack saddle up underneath one of the now vacant lifting points.

> *Top Tip* Be super aware of how you position the Trolley Jack; see the Top Tip in Chapter 5 on Page 78 regarding when the car has been lowered.

When you're happy the Trolley Jack saddle is correctly aligned, carefully and slowly start to lift the car clear of your primary Axle Stand; keep a keen eye on the jack saddle's contact with the lifting point as you lift. If the saddle starts to slip - STOP immediately, lower the car and reposition the jack saddle.

You don't need to lift the car very high, just high enough to be able to slide the Axle Stand out and well clear of the vehicle. When the Axle Stand is well clear, slowly lower the car back onto the ground, then move your trolley jack away from the vehicle out of harm's way. Chock both sides of the wheel that's now back on the ground and repeat the process on the other side of the car.

Complete Front Wheel Bolts Tightening

Step 5: To complete tightening the wheel bolts, for most Porsche models, set your torque wrench to **130Nm** (96ft-lb) (check your handbook for your correct torque setting). Take your Locking Wheel Nut Key and Torque Wrench and check that your Torque Wrench is set to *tighten*. Starting from approximately the 12 o'clock position, torque down all five bolts in the usual *star* formation. Be careful not to hit the Torque Wrench handle against the side of the car. When all five bolts are tightened to 130Nm, *(or your correct torque setting)*, finish by applying a small squirt of WD-40 to the bolt heads to help protect against corrosion.

> *Top Tip* This safety tip comes courtesy of Golding Barn Garage. See the Top Tip in Chapter 5 on Page 78 regarding wheel bolt torques.

Day 2

Stages 1-8: Rear Alloys, Calipers & Tips Decontamination

When you've replaced the magnificently ceramic coated front wheels on the car, turn your attention to the rear wheels.

Chock the Wheels and Loosen the Nuts

Follow the same steps explained in Step 2 on Page 68, this time chocking the front wheels with two chocks per wheel.

Raising the Rear of the Car: When you've replaced the front wheels on the car, move to the rear and jack up the rear end in a similar way. Remember to chock the front and back of each front wheel first, once again with two chocks per wheel. If you have access to a Long Reach Trolley Jack, lifting the rear is a bit easier than the front.

I like to jack up the rear using the rear suspension mounting point located at the mid-way point of the undercarriage, between the two rear wheels. This spot attaches the suspension to the rear chassis and is easily strong enough to lift.

Jacking the car up from this position will usually raise both rear wheels off the ground together. This is ideal as it enables you to set both Axle Stands to the same height and put them carefully and securely in place on both sides of the car under the factory standard jacking points at the same time. *Do not* attempt to lift the car using any part of the engine or gearbox. Slowly, carefully lower the vehicle onto the axle stands and remove the Trolley Jack away from the car out of your way.

Check the Car's Stability

Follow the same process explained in Step 4 on Page 70.

Rear Wheels Removal

Follow the same steps explained in Step 5 on Page 71.

Exhaust Tips TLC

Having removed the two rear wheels, once again complete the wash and decontamination (Stages 1-7) on the rear wheels, calipers, and this time *also* on the exhaust tips (missing out Stage 7 *'To Polish or Not'* on the exhaust tips). Don't neglect the inside of the tips; your Meguiar's Supreme Brush should get right in there nicely.

For those of you who have *chrome-finished* exhaust tips that may have seen better days and are rather tired looking and lacking *lustre,* here's how to bring them *galloping back* to life before ceramic coating and locking in the finish.

So, with Stages 1-6 done on your tips, dry off your tips with a clean Microfibre Cloth. Then, take your Peek Metal Polish, Extra Fine Steel Wire Wool, Microfibre Applicator Pad and a clean, dry Microfibre Cloth. Tear off a small wad of wire wool, apply a pea-sized blob of Peek to the wire wool and work it briskly over the entire exhaust tip for a minute, including the front lip and underneath bits that you can't see. You'll see the Polish start to blacken, but that's normal; it's just the paste absorbing the dirt. Then, *furiously* buff off the polish with a clean Microfibre Cloth, folding it as you go. Next, apply another blob of Peek, this time to your Microfibre Applicator Pad. Once again, work the Polish briskly all over the tip before giving it a final frenzied buffing with a clean Microfibre Cloth. When finished, move your Metal Polish, Wire Wool, Applicator Pad and used Cloth away from the car out of the way.

Stand back and check your results; I think you'll be flipping delighted! I recently detailed a rather nice 2014 Golf GTI, and I did this process on the pitted, drab-looking chrome tailpipes. The owner was so amazed by the stunning transformation of his tips that he thought I'd replaced them with new items! Happy customer = happy Driveway Warrior – brilliant!

Complete your tips by giving them the Stage 8 IPA wipe down in preparation for ceramic coating.

C5 Wheel Armour Application – 5 Steps

Finally, you can complete the above-described *5 Steps* to apply C5 to your two rear calipers, exhaust tips, and rear wheels. To avoid cross-contamination, change Mini Applicator Pads each time you go from calipers to tips to wheels. Allow the rear wheels to cure for 12 hours before replacing them on the car.

Get Snappy

You are DONE; congratulations! Stand back and admire your newly transformed, gleaming *holy trinity*. If you fancy taking some shots to capture the moment of wheels, calipers and tips detailing perfection, I suggest you have a peek back toward the end of Chapter 1. I provide hints and tips for achieving cracking driveway photos on your smartphone.

5 Steps to Safely Jacking Up the Car to Replace the Rear Wheels

At the end of day 2, your fabulous newly ceramic coated rear wheels will be ready to go back on the car. Repeat Steps 1-5 described on Pages 99-101 to replace the rear wheels on the car.

Clear Up

I know, clear up is boring; you're probably tired and can't be bothered. As I keep banging on, clear up is a necessary evil and will ensure all your precious detailing and maintenance kit stays in *tip-top* condition, ready to go next time you need it. Start by returning all your Liquids and Spray bottles to storage, but do check all lids and spray heads are secure and keep an eye on any dilutables that need topping up. If you used the Handheld Foam Sprayer, rinse it out with clean water, semi-pressurise it, and spray out some clean water to clear the feed pipe and nozzle. Give it a quick dry and store it away.

Grab your Car Blower, Caddy, Mats, Stool, Detail Guardz and Wheel Chocks. Give them a quick wipe down to dry if necessary and store them away. Give your

Trolley Jack and Axle Stands a quick wipe down with WD-40 and put them safely away.

Return your Locking Wheel Nut key to its place of safekeeping. Release the tension on your Torque Wrench and store it in its box (never keep the Torque Wrench under tension as this will reduce the tool's life).

Discard all the bucket water and give your Wheel Bucket and Grit Guard a rinsing blast with the Pressure Washer or Hose Nozzle but hold back the Wheel Bucket for now as you'll need it a bit later. Next, gather all your used Cloths and Mitt and squeeze them out if necessary. The Microfibre Cloths will be mucky from wheel duty, so chuck them in your Wheel Bucket, half-fill it with clean, warm water, add a small squirt of washing-up liquid and give them a short pre-wash with your gloved hand.

Then squeeze them out, discard the dirty wash water, give the Wheel Bucket a quick rinse out, and pop the Cloths in the washing machine at 40°C (104°F). Add the Microfibre Detergent and stick on a 30-minute cycle with a gentle spin.

When you take them out of the machine, give them a good shake to re-set the fibres and hang everything on a clothes dryer to dry naturally, using a peg for the Wheel Mitt.

You should be left with just your Pressure Washer, Wheel Brushes and Wheel Bucket. Dump all the Brushes in the Wheel Bucket and give them a quick rinsing blast with the Pressure Washer or Hose Nozzle. Discard that rinse water, add a squirt of washing-up liquid to the bucket and half-fill it with hot tap water, swirling it around to mix.

Spend a couple of minutes with your gloved hands in the soapy water, massaging the bristles clean. Then dump the dirty soapy water and give the Brushes a final rinsing blast in the bucket with the Pressure Washer or Hose Nozzle and dump that rinse water in the water butt.

Give your Wheel Bucket a quick rinse, wipe it dry and store it away with the clean Grit Guard. I then give all the Brushes a good wrist flick to fling off the excess rinse water, and I lay them on an old towel under the radiator to dry or in the sun if you're lucky. Give your Pressure Washer and accessories, including the power cable, a wipe down to dry and store them away. The next day when the Cloths,

Mitt and Brushes are fully dry, put them away, keeping your Wheel Brushes in your Wheel Wash Bucket.

Reflection

You've done a brilliant job completing your wheels, calipers, and tips ceramic coating project. Super well done, YOU! Oh, the sense of satisfaction at achieving up to 24 months of protection. You'll still have to clean them, of course, but you should find that a blast with the Pressure Washer or Hose Nozzle during your regular maintenance safe wash will be enough to keep them gleaming and in tip-top nick for the foreseeable.

Enjoy some downtime, knockback those beers on ice, and reflect on the £291 ($392) (the average of three quotes received) saved on what the pro detailer would have charged you for the job you've so brilliantly done yourself - you're a star, Driveway Warrior!

Downtime complete, have you Porsche Cayman owners ever wondered what lies beneath your engine cover? I can tell you it's probably a pretty filthy, shamefully never detailed engine, and the same goes for all you other sports car-owning Driveway Warriors. By now, you may have detailed to the *nth degree everything* on the car you can actually see, so maybe it's time to turn your attention to the areas you *can't* see, like the engine. They still need TLC, so let me show you how with an *Exquisite* Engine Detail and Protection Project - Driveway Detailing Warrior *style,* no less. Read on; in Chapter 7, I'll explain precisely how …

Chapter 7

Exquisite Engine Detail and Protection

Sunday 8 am. It's a glorious spring morning; I'm out for an *invigorating* early morning blast in the Cayman, *tearing* along my deserted favourite Sussex A-Road, approaching *warp* speed (within the speed limit, naturally). The Porsche feels planted like a rock, just how she's supposed to feel. I tramp the clutch and *smoothly swipe* the gear stick, downshifting from fifth to third, grinning from ear to ear as the revs *soar* and the performance tweaked 300 bhp *flat-six* power-plant *howls* like a demented banshee. I'm pinned back in my seat, tightly grasping the wheel as the Cayman *catapults* forward. But wait, a random thought hits me like a *juggernaut*. My 11-year-old mid-engine power plant *sounds* utterly *awesome,* but does it *look anywhere* near as good as it sounds? I must investigate this intriguing dilemma at once - read on, friends and let's find out…

Back on the safety of my driveway, I'm watching a quick YouTube video on my phone about how to access the covered mid-engine. A brief wrestle to remove the carpet sound-proofing and the metal engine cover comes off quickly enough. *One* filthy and disgusting *Direct Fuel Injection Type MA1.20 2.9 Litre* power plant is revealed there. My heart briefly sinks; it's clearly never been cleaned or detailed before. My mission is clear – bring the engine up to standard so it matches the magnificent aural *throaty growl* when under load. Let's do this, friends: Driveway Warrior fashion, of course …

The Natural Elements

Let me refer you to Chapter 1 on Page 2, 'Sunshine – the Safe Wash Enemy.' All the advice given in previous chapters about detailing in ideal weather conditions applies to your Exquisite Engine Detail and Protection Project. Never start or finish your Detailing Processes in direct sunlight if you want to avoid issues with your detailing liquids drying out too quickly on your car and screwing up the quality of your finish. Pick a day when the forecast is dry and mild, preferably overcast, when your paintwork is cool to the touch. We want to give ourselves every opportunity to achieve the perfect results, right?

While you're checking the weather forecast, look if it's going to be windy. If you're detailing a Porsche Cayman or similar mid-engine vehicle, *don't* make the same mistake I made by working with the protective Polythene Dust Sheet in the wind. It becomes a total nightmare to lay the sheet down, which will slow you down and frustrate you no end – who needs that if it can be avoided?!

Exquisite Engine Detail & Protection Kit List

The kit for this project is much the same as for the *Wheels off Wheel Arch Detail* in Chapter 5, with one or two changes. Notably, you won't need the pressure washer on this one.

Detailing Hardware Required:

- Carry Caddy
- Nitrile Gloves
- Hosepipe & Nozzle
- Wheel Wash Bucket
- Grit Guard & Wheel Wash Kit
- Car Blower & Extension Lead
- Detail Guardz
- Handheld Foam Sprayer (diluted 1:5 APC to De-ionised Water)
- Microfibre Cloths (a bag full of downgraded cloths)
- Eye Protection
- Face Mask
- Polythene Dust Sheet 3.5m x 2.6m (12ft x 9ft)
- 3M Auto Detailing Masking Tape
- Sharp Scissors
- *Grabber Tool

*This long reach tool is perfect for retrieving anything dropped into hard to reached spots, e.g., when detailing your engine bay. It is available in different lengths and cheap as chips - you need one in your detailing kit!

Detailing Liquids Needed:

- Rust Repellent (diluted 1:3)
- Pure Shampoo
- Perl Dressing Solution Spray (diluted 1:2)
- All-Purpose Cleaner (APC) Spray (diluted 1:5)
- Isopropyl Alcohol Spray (IPA) (diluted 70:30)
- Fallout Remover Spray
- Tar & Glue Remover Spray
- Microfibre Detergent

Engine Cover Removal & Replacement Tools

- Plastic Trim Removal Tools
- T30 Torx Bit
- Socket Spanner/Wrench 6.35mm (¼in)
- Container for Bolts

A Word on Engine types and Layouts for Detailing Purposes

Frankly, the mid-engine layout of the Cayman makes it an absolute pain in the proverbial to detail. The engine bay is surrounded by carpeted areas of the rear cargo bay, so care is needed when using the hosepipe and detailing products in this area. But, where there's a *will*, there's a *way*, right? The good news is that *your* engine layout may be much easier to detail. If you have a 911-style rear engine or a traditional front-loaded engine, they all lend themselves to straightforward detailing. And the techniques I describe in this project are transferable to *any* modern engine.

Is it Safe to Detail the Engine?

I know some of you may have concerns over the safety of detailing your engine bay for fear of damaging something in there. But fear not, *modern* engines are generally watertight, and with a bit of care and attention to detail, you won't harm or break anything in your engine. I skipped the pressure washer for this project, as I don't want to blast any delicate electricals with the powerful pressurised spray. So, I stick to the Hosepipe Nozzle to rinse down the cleaning stages, as the gentle free-flowing shower spray won't hurt anything in the engine – safe as houses!

Note that I said *modern* engines. I've never detailed a classic car engine, so I can't speak for how watertight they may be; I suspect not very. If you have a classic car, please proceed with extreme caution.

Why Bother? I Never Look at My Engine

Hold on a second, my friend, we Driveway Detailing Warriors *always* bother - you know it's all in the *detailing*, right? It's not *just* about aesthetics either; a clean and detailed engine is safer and easier to work on. For starters, you're less likely to get your hands and clothes covered in oil and grease, and it's *much* easier to spot trouble early, e.g., any coolant or oil leaks on a detailed engine will stick out like a sore thumb.

It's not so easy to spot fluid leaks on an engine covered in grease and grime - *how would* you know where it's coming from and if the leak is old or new? It's also true that oil and grease can *accelerate* wear on rubber hoses and plastic parts - so detailing it *will absolutely* help prolong it!

If you don't service your car yourself like most people, your mechanic will also thank you for making their job easier when it comes to servicing the engine. As for the aesthetics, well, I guarantee it will make you happy knowing your engine looks as good as it sounds. Plus, everyone knows a detailed engine adds at least *20bhp* in the same way that a go faster bonnet stripe does, right?

Here's a short tale about the cosmetic appeal of a clean engine while selling my much-loved GT Silver Porsche 911 996 model a few years ago. Upon opening the rear lid to show the prospective buyer the freshly detailed, *gleaming* engine and bay, he was so astonished he asked me if the 88K miles engine was new. He commented he could *eat* his dinner off it, and I genuinely think it helped make his mind up. Well, that and a *thrilling rapid* test drive blast over the Sussex Downs meant a sweet deal was reached!

Preparation

Remember to get those beers in the fridge, ready for when you finish and grab a bottle of water or Thermocafé mug of coffee to sustain you.

Remove your rings and watch, and make sure you aren't wearing anything zippy or with buttons or poppers that could scratch your car while working up close to it. Position your vehicle preferably with room to safely work around it. Place your Detail Guardz under the outer edges of the rear tyres. Get your Engine Detailing & Protection Kit and Tools assembled. Load up your Carry Caddy with your Spray Potions, Microfibre Cloths, Foam Sprayer, Trim Tools, Grabber Tool, etc., and have your Caddy handy.

Before you begin, it's essential to make sure the engine isn't too hot. We all know by now that detailing products don't react well to hot surfaces. But, there's also a safety issue here; you obviously don't want to burn yourself. I've found that a *slightly* warm engine helps to more quickly *banish* the more stubborn areas of built-up grease and grime. So, after the engine cover is removed, if the engine is cold, simply start your car up and run the engine for **two minutes only**, then turn the engine off.

> ***Top Tip*** The Porsche Cayman rear hatch has *two* open positions: the first day-to-day position, and then if you push the hatch past that first position, it will go up to the second *service* position, making it easier to access and work in the engine bay.

Remove the Engine Covers

This step is for you mid-engine Porsche Cayman owners; your engine lives under a metal cover, beneath some carpeted sound deadening material, behind the two seats. Here's how to remove everything to access the engine; it's pretty straightforward.

All you non-mid-engine car-owning Driveway Warriors can skip this step.

> ***Top Tip*** Don't use a metal screwdriver or anything else metal to remove the carpeted sound-proofing, as you may damage the rubberised type trim finish around the engine bay. Your Plastic Trim Removal Tools are perfect for this task.

Improve the Access

To improve access to the engine bay from inside the car, push the seats as far forward as they'll go. Remove the cargo net *(I wish I could find a red cargo net)* from the top of the carpeted cover and put it somewhere for safekeeping.

Wearing your Gloves, take a couple of your wide flat-ended Plastic Trim Removal Tools, and insert the flat ends just inside the two cargo net retaining clips - between the carpet cover and the rubberised trim behind the seats. Gently lift the carpet cover up; depending on how many times it's been removed before, the carpet cover will come up easily, or it may need a bit of *finessing,* but not too much. You may need to go around the other side of the car to work the second Trim Removal Tool. When you have both sides of the carpet cover lifted, slide it back towards the seats, and the two retaining clips on the front of the carpet cover will come clear.

The carpet cover is surprisingly heavy due to all the inbuilt sound deadening material; remove it from the car and place it somewhere for safekeeping.

Depending on what year your Cayman is, there may be a long piece of foam rubber tucked in at the front edge of the metal engine cover. Just remove it and place it somewhere safe.

Untighten the Torx Bolts

The metal engine cover will now be exposed. Five Torx bolts hold the metal cover secure – one near each corner and one on the back edge in the middle.

Take your Socket Wrench and T30 Torx Bit and untighten the five bolts - you *don't* have to loosen or remove them in any particular order. The five bolts are all exactly the same, so you don't need to worry about putting them back in the same order. Remove the five bolts and put them in your handy container away from the car, somewhere safe.

The next step is to remove the metal engine cover; it has a rubber seal around the underside edge. Depending on how frequently your engine has been serviced, the cover will come off easily, or it may prove a bit stuck.

If it is stuck, grab one of your trusty Plastic Trim Tools and ease it under one corner at the front, give it a wiggle, and the cover will soon come free. Ease it up and out of the car without *smacking* the edge against any interior trim, causing a filthy scuff! Place the metal cover with the bolts and carpet cover for safekeeping.

There She Is - the Mucky Devil!

So, that's what your Cayman power plant looks like!

Now, you can assess how dirty the engine bay is and crack on with the detailing, but first, cover up and protect all your lovely carpeted rear cargo bay and seats.

At this point, I need to quickly circle back to Page 111, where I advised I've found that a *slightly* warm engine helps to *banish* the more stubborn areas of built-up grease and grime quicker. So, now the engine cover is removed, simply start your car up and run the engine for *two minutes only*, then turn it off.

Cover the Cayman Carpeted Rear Cargo Space

Ok, I'm not going to pretend this isn't anything other than a flipping fiddly and rather irritating but *essential* step for Porsche Cayman owners. You must do it to protect the seats and carpeted rear interior areas from splashes and drips from the hosepipe nozzle and your detailing products. All you non-mid-engine car-owning Driveway Warriors can blessedly skip this step – lucky you!

Clean It so You Can Stick It

The first step is to take your IPA Spray bottle and spritz some IPA right around the edges of the engine bay, where you'll be sticking the plastic sheeting down. Then wipe the area down with a clean, dry Microfibre Cloth; this is going to hugely help your masking tape stick properly where you want it to stick.

Top Tip I prefer the sky-blue variety of 3M auto masking tape; it's comparatively reasonably priced and widely available in different widths. I use the 19mm stuff because it's easy to curve around the contours of your car and is wide enough to cover a strip of rubber trim. Its low-tack rubber-derived adhesive tends not to lift even when wet or leave behind any messy residue.

Lay Down the Dust Sheet

Next, with your Gloves off, take the large piece of Polythene Dust Sheet and spread it out over the whole rear cargo and engine bay area. Let it droop down loosely over the engine bay from the top, near the back of the roof - now you can see why it's impossible to do this if it's windy!

Cut-Stick-Press

Once you have the sheeting roughly laid out in place, hanging down loosely from the top near the back of the roof, tape the top of the sheet to the top of the hatch opening. At this point, you may find access a bit easier by pushing the sheeting forward and climbing into the cargo area.

Cut It: Now, take your sharp scissors and carefully cut around the edge of the engine bay aperture. You want to try and cut the engine access hole a little too small, as this will enable you to tape down the excess plastic to the edges of the engine bay. When finished cutting, move the scissors out of the way.

Stick It: Then, using *small* pieces of overlapping tape, begin taping the sheeting down to the metal back edge at the top of the engine bay. There's no other way to say it; this is slow and tedious work, but tell yourself you'll probably never have to do it again. Persevere, and you'll find your rhythm and start to get faster. Work your way around the whole oval-shaped engine bay opening until the sheeting is secured right around the edge with slightly overlapping small pieces of tape.

Press It Down: Go around the tape, pressing it down with your thumb, making sure it's good and stuck. When the engine bay is taped down, you can arrange the rest of the sheeting on all four sides of the engine bay. It needs to cover the entire rear cargo area and go up the back end of the engine bay up to the top of the hatch opening so that no rinse water can enter where the seats are. Finally, arrange the sheeting so it goes up either side of the engine bay, left and right over the sides of the rear cargo area, and hangs down the exterior sides of the car.

At this stage, have a quick check that there aren't any tears in the sheeting; if there is, no problem, just put some tape over the tear, and that should be good to go – phew! This effort should successfully keep rinse water on the *engine* and *away* from anywhere else that shouldn't get wet.

8 Stages to Exquisite Engine Detailing & Protection

Grab your Caddy loaded with your detailing products. Take your Wheel Wash Bucket, check your Grit Guard is in the bottom of the Bucket and add two Pure Shampoo capfuls, plus a capful of neat APC for extra foaming and cleaning power, and fill with warm water. Swirl the water around to mix the shampoo and APC.

Stage 1

Rust Repellent: With your Gloves on, take your Rust Repellent Spray and shake it well. For a mid- or rear-engine car, give the rear brake discs (rotors) a good spray down with Rust Repellent and allow to cure for a couple of minutes. Do the same to the front discs (rotors) for a front-engine car. You need to do this step because you'll be working with the hosepipe, rinsing down the engine on several of the steps, and some rinse water will inevitably get on the brake assembly. Your Rust Repellent treatment will minimise the dreaded orange hue settling on your discs as soon as they get wet.

When finished, move your Rust Repellent spray away from the car out of the way.

> *Top Tip* Don't forget to roll the car a few inches to access the bit of the disc (rotor) hiding behind the brake caliper.

Stage 2

Car Blower: You're now working with the engine close to eye level, so please put your eye protection on. I like to start the engine work by giving the whole engine bay area a thorough air blast with the Car Blower to dislodge any loose dirt particles and blast them away. Don't forget to hook the power cable over your shoulder so it doesn't trail scratchily over the paintwork. When finished, move your Car Blower away from the car out of the way.

Stage 3

Pre-Soak with APC Foam: Grab your Handheld Foam Sprayer loaded with a 1:5 mix (APC to De-ionised Water), shake well, and liberally foam up *everywhere* in the engine bay. Allow the foam to sit for 3 minutes, then give the whole engine bay a thorough rinse down with your Hosepipe Nozzle set on the shower spray.

Stage 4

Two-Step Brush Attack with APC Foam

First Step: Now, generously foam up the entire engine bay with the APC Foam Sprayer for the second time and allow the foam to soak for two minutes. Then, take your Wheel Wash Bucket & Wheel Wash Kit and get scrubbing everywhere!

Start with the Meguiar's Supreme Brush, regularly rinsing it in your Wheel Wash Bucket and work your way through your Wheel Brush Kit. Keep them super-well-lubricated with wash water as you go (a dry brush is a scratchy one).

Use the longer handle Brushes to reach down into the depths to tackle there as best you can. Feel free to turbocharge your Brushes with a few squirts of APC Spray (not the foam) as you go. Use your Wheel Bucket Wash Mitt to massage over the engine parts your hand will reach, in conjunction with all your Brushes. When finished with the Brushes on this first step, give the whole engine bay another thorough rinse down with your Hosepipe Nozzle.

> *Top Tip* If you drop a small Brush or Plastic Trim Tool down into the engine bay depths, no problem, take your Long-Reach Grabber Tool and snag the offending item out of harm's way.

Second Step: Next, generously foam up the entire engine bay with the APC Foam Sprayer for the third and final time and allow the foam to dwell for another two minutes. You should just about get three applications from one full bottle of foam mixture. Then repeat the above Brush and Wash Mitt attack and finish by giving the whole engine bay another thorough rinse down with the Hosepipe Nozzle. Move your Handheld Foam Sprayer away from the car out of your way.

Check how dirty your Wheel Wash Bucket water has become, as it's probably pretty nasty by now. Discard it, rinse the Bucket, and refresh it with another bucket of clean, warm water mixed with two capfuls of Pure Shampoo.

Stage 5

Brush Assault with APC Spray (non-foaming): Grab your APC Spray bottle mixed at 1:5 and shake well. Generously spray down the whole engine bay with non-foaming APC, including the areas you can't see, and allow it to soak for two minutes. Take your ValetPro Chemical Resistant Wheel Brush and work the Brush over the whole area, rinsing and lubricating it regularly in your Wheel Wash Bucket as you go.

Utilise your best-matched Wheel Wash Brushes to reach in the nooks and crannies as best you can. When finished with the Brushes on this stage, give the whole area another thorough rinse down with the Hosepipe Nozzle. Move your APC Spray away from the car out of the way.

Stage 6

Brush Barrage with Tar & Glue Remover: Your engine will be starting to *gleam* by now, but you can't relax yet. When you drive your car, your engine is subject to attack from tar spots from underneath, which will *relentlessly* find their way all over it. So, take your Tar & Glue Remover Spray, shake well, and generously spray down the whole engine bay, allowing it to soak for two minutes. Then, take your ValetPro Chemical Resistant Wheel Brush and work it over the entire area, rinsing and lubricating the Brush regularly in your Wheel Wash Bucket as you go. Utilise your best-matched Wheel Wash Brushes to reach in the nooks and crannies as best you can. When finished with the brushes on this stage, give the whole engine bay another thorough rinse down with the hosepipe nozzle. Move your Tar & Glue Remover Spray away from the car out of the way.

Stage 7

IPA Wipe Down: We're at the final cleaning stage, phew! Grab your Car Blower, hook the power cable over your shoulder so it doesn't trail scratchily over the paintwork, and spend a few minutes giving the engine bay a thorough air blast to expel most of the rinse water out of the nooks and crannies. Then, to prepare the engine to receive the dressing solution/protectant, take your IPA Spray diluted 70:30, shake well and spritz down the entire engine bay, getting the Spray into all the tight spots. Next, grab your bag of downgraded Microfibre Cloths, and wipe down the whole engine. Dry it as much as you can, folding the Cloth as you go and changing to a dry Cloth when necessary. You may find it helps to grab one of your longer Plastic Trim Removal Tools, wrap your drying Cloth around one end and use that Tool to delve into the depths to dry hard-to-reach areas.

You should find these seven *relentless* detailing stages thus far have been successful in cleaning the whole engine bay beautifully. Your Microfibre Drying Cloth should be coming up satisfyingly clean – most likely, the *cleanest* the engine has been since the car left the factory! Move your IPA Spray away from the car out of the way.

At this point, I like to start the engine to warm it up for a further *three minutes only*. There will be small pools of rinse water lying in the depths that you can't reach with your Car Blower or drying Cloth. Once the engine is *warm, not hot*, any remaining water will quickly evaporate. When you start the engine, check if you have any new warning lights come on, which I have never had, but it's still worth checking.

Stage 8

Lay Down Luscious Dressing/Protectant: Now for the really satisfying bit; this is my favourite part of the detailing process. By now, your engine should be slightly warm to the touch, *not* hot, and be dry enough to dress. *Carefully* check how warm the engine is before you touch it again. If it's good to go, take your CarPro Perl Dressing Spray, diluted 1:2, shake well and mist the Perl over the entire engine, concentrating on all the plastics and rubbers. Allow it to soak for two minutes, then work the Perl in well using a fresh, downgraded Microfibre Cloth. Fold the Cloth as you work the dressing into the plastics, rubbers, ribbed piping and hoses as best you can. At this point, you may find it helps to get into those narrow areas by taking a flat-ended Plastic Trim Tool, wrapping the Cloth around the flat end, and using that to work the dressing into the tight spots.

As well as the *magnificent restorative*, mid-sheen darkening of the finish, the Perl also gives the plastics some welcome UV and weathering protection. Due to the amount of abuse the engine bay receives from day-to-day driving on our lovely Autumn and Winter roads, I recommend doing two coats of dressing in the engine bay.

Allow the first coat of dressing to cure for ten minutes, then repeat the process for the second coat. You'll notice an even deeper darkening and sheen of the plastics and rubbers after the second coat – lovely! Move your Perl Dressing Spray away from the car out of the way.

Get Snapping

You are DONE; congratulations! Stand back and admire your newly transformed, gleaming engine, now befitting of the *mighty* badass aural *snarl* of the power plant under load! If you fancy taking some shots to capture the moment of detailed engine perfection, I suggest you have a peek back toward the end of Chapter 1, where I provide hints and tips for achieving cracking driveway photos on your smartphone.

Replace the Engine Cover & Sound Proofing

Spruce Up the Metal Cover: First up, remove all the plastic covering and securing tape and bin them. Check if any areas where the tape has been removed need wiping dry. Before the metal engine cover goes back on, you need to give it a quick spruce up, particularly the underside. You don't want a mucky cover going over your newly mint engine now, do you? Grab your APC Spray, spritz down the metal cover, including the rubber seal on the underside and allow it to soak for a minute.

Take a clean Microfibre Cloth and wipe everything down. Check if it needs a second hit with APC. If it looks good, take your IPA Spray and spritz down the rubber seal on the underside, leave it to soak for a minute, then wipe the seal dry with a clean Microfibre Cloth – this will prepare the rubber seal to receive some dressing.

Dress the Rubber Seal: Finally, grab your Perl Dressing Spray and spritz right around the rubber seal; leave it to soak for a minute. Then work the dressing in with a clean Microfibre Cloth, folding the Cloth as you go. Allow to cure for five minutes, then apply a second coat of Perl and work it in again with the Cloth. The Perl will nourish and protect the rubber, prolonging its life.

If you removed a long piece of foam rubber tucked down at the front edge of the engine bay, go ahead and return it. Now you can replace the metal cover and secure it with the five Torx bolts. There isn't a torque setting for these bolts specified in the Porsche handbook, so don't overtighten them. Next, go ahead and replace the carpeted cover; it goes in the same way it came out; just line up the tabs and push it into place – you may need the Plastic Trim Tool to help you snug it into place. Finally, replace the cargo net, and you are done, my friend – congratulations, *YOU!*

Clear Up

It doesn't matter how many times I say it; clear up will remain in equal measure, a pain in the posterior and utterly *crucial* to keeping your precious kit in tip-top nick, ready to go next time you need it.

Start by returning all your Liquids and Spray bottles to storage, but do check all lids and spray heads are secure and keep an eye on any dilutables that need topping up. Retrieve the Handheld Foam Sprayer, rinse it out with clean water, semi-pressurise it, and spray out some clean water to clear the feed pipe and nozzle.

Give it a quick dry and store it away.

Grab your Car Blower, Extension Cable, Caddy, Detail Guardz and Grabber Tool, give them a quick wipe down to dry if necessary and store them away.

Discard all the bucket water and give your Wheel Bucket and Grit Guard a rinsing blast with the Hose Nozzle but hold back the Wheel Bucket for now as you'll need it a bit later. Next, gather all your used Cloths and Mitt and squeeze them out if necessary.

The Cloths are likely to be mucky from engine duty, so chuck them in your Wheel Bucket, half-fill it with clean, warm water, add a small squirt of washing-up liquid and give them a short pre-wash with your gloved hand. Then squeeze them out, discard the dirty wash water and give the Wheel Bucket and Grit Guard a quick rinse.

Pop the Cloths and Mitt in the washing machine at 40°C (104°F), not forgetting to add Microfibre Detergent and stick on a 30-minute cycle with a gentle spin. When you take them out of the machine, give them a good shake to re-set the fibres, and hang everything on a clothes dryer to dry naturally, using a peg for the Mitt.

You should now be left with just your Hosepipe and Nozzle, Wheel Brushes, and Wheel Wash Bucket. Dump all your Brushes in the Wheel Bucket and give them a quick rinsing blast with the Hose Nozzle. Discard that rinse water in the water butt if you have one, add a squirt of washing-up liquid to the Bucket, and half fill it with hot tap water, swirling it around to mix. Spend a couple of minutes with your gloved hands in the soapy water, massaging the Brushes clean. Dump the dirty soapy water, give the brushes a final rinsing blast in the Bucket with the Hose Nozzle and then discard that rinse water in the water butt.

Give your Wheel Bucket a quick rinse if needed, wipe it dry and store it away with the clean Grit Guard. I then give all the Brushes a good wrist flick to fling off some of the excess rinse water, and I lay them on an old dry towel under the radiator to dry, or in the sun if you're lucky. Give your Hosepipe and Nozzle a wipe down to dry, and store them away. I don't keep my *Hozelock* branded nozzle out in the elements; instead, I keep it safely in my shed. The next day when your Cloths, mitt, and brushes are fully dry, store them away, keeping your wheel brushes and wheel mitt in your wheel wash bucket with the grit guard.

Reflection

Here's a brief thought about that black cargo net in the Cayman rear cargo area. I've been hunting high and low for a *red* version to match the red interior trim. It would offset the stone-grey interior beautifully as a nifty little mod, but alas, I can't find one anywhere. I wonder if that nylon net material could be successfully dyed red – hmm, I somehow doubt it! If any Driveway Warriors have any ideas about getting a red one, please let me know.

Now your engine looks as *breathtakingly* good as it sounds; what a flipping *fantastic* feeling that is, Driveway Warrior. As I jokingly mentioned earlier, your power plant will have increased at least 20bhp now. I swear to the almighty, it sounds blinking *better* than ever - or is that just my ears playing tricks on my brain? Oh, who cares!? Time to chill, enjoy those icy beers, and reflect on a job *awesomely* done.

Oh, and there's the small matter of £238 ($319) saved on the pro detailer's quote for the project you've completed yourself so *brilliantly*.

But wait, you've detailed everything you can see on your car, even the engine that you can't see! What the hell else can their flipping well be? Perhaps your thoughts will drift now to that desired paint correction/restoration project, the one the professional detailer quoted far more than you wanted to pay. Read on as I reveal how YOU can do a fantastic, safe paint correction/restoration, Driveway Warrior style, of course …

CHAPTER 8

Perfect Paint Correction/Restoration & Protection: Methods for Hand & Budget Machine Polishing

Part 1 - The Paint Correction Decision

The used car salesman looked at me coldly. He'd just appraised my second-hand seven-year-old VW Golf and said, "Your car looks as though it's been cleaned with a scouring pad; it's crying out for paint correction. I'll give you £900 ($1214) to take it off your hands."

This *humiliating, depressing* experience happened to me many years ago when I was looking to sell my car. The Golf's true value was £5000 ($6747), so that was the day I started my journey into the world of paint correction and restoration!

Before you go and unleash your serious kit like Clay Mitt, Cutting Compound, Machine Polisher, etc., you should ask yourself one fundamental question. Does my paintwork really *need* correction/restoration? Funnily enough, the answer isn't always completely obvious, particularly when our judgment may be clouded by our wants and desires rather than by what our paintwork *needs*.

What are Swirl Marks?

It seems that people in the detailing world use the term 'swirls' to describe a variety of paintwork problems. Fundamentally, swirl marks are the marring visible due to light reflecting tiny scratches on your car paintwork. So, yes, swirls are microscopic scratches in the clear coat. I consider them to be the enemy of your clear coat as they can deteriorate over time if not corrected in a timely way.

There are many ways to introduce swirls to your paint; that's why we detailers steer well clear of automatic car washes (aka swirl factories or scratch washes). We don't wipe down our dusty car with any manner of dry cloth, and we even rip off the labels before touching the paint with a new microfibre towel! The safe wash techniques explained in Chapter 1 go a long way to minimising the introduction of swirls in your paint. However, unless you bought the car from new, some previous, uninformed owner saboteur may have been merrily scrubbing away at the bodywork with a scratchy chamois leather or an old t-shirt!

Stick or Twist

You're the Driveway Detailing Warrior, and it's your paintwork and decision to make, not to mention your time, effort, resources, and detailing products. Don't let your mates, neighbour, postman, or even your Porsche indie mechanic (unless they're also a bodywork specialist) tell you what your paintwork needs. You'll need to be super unbiased in making your decision; trust your senses and a healthy dose of your common sense, and you'll come to a rock-solid decision to stick or twist. Of course, if you remain unsure, there's no harm in asking someone you trust for a second opinion.

Having completed your safe wash and the paintwork is dry but un-waxed (see Chapter 1 after Step 10A), take several steps back and slowly walk around the car, running your eyes over the paintwork. View it from different angles and distances; what do your eyes tell you, is the paint a bit tired and faded? Are there numerous small but annoying swirls and scratches marring the surface? Lightly run your (un-gloved) fingertips over the paintwork at different points, including along the lower third of the bodywork. Does the paint feel a bit rough or gritty, almost like fine sandpaper?

If the answer is mostly yes, then your paintwork is contaminated and likely does need correction/restoration. Contaminants comprise many particles harmful to your paint, such as road tar, tree sap, industrial fallout, and even iron filing particles from train tracks.

Unfortunately, safe wash alone will not remove these nasties, but fear not, you CAN do this. Here's how you can also avoid the punishing fee quoted by pro detailers to do the job.

Machine v Hand

Now for a few thoughts on cheap machine polishers' merits against hand applicator polishing. A pro detailer will wax lyrical about the need for a £400 ($527) sexy branded Dual Action Polisher, and I have no problem with that approach; it's just not for me. Believe me; the Driveway Warrior can achieve authentic, spectacular results by correctly using a budget machine polisher as part of a structured paint correction process or by using a humble hand applicator, including impressive bicep development with the latter!

My trusty *Silverline orbital machine* cost just £19 ($25) new from eBay, though I grant you it does make a frightful racket when cranked up! Having trialled both machine and hand applicator methods on my Cayman, I can say with authority that I could not tell the difference between the two finishes, even under the glare of a fearsomely bright LED light. However, the clear advantage of the machine polisher was the speed with which it got the job done, taking about half the time of the modest hand applicator.

Those Pesky Amateurs

It's reasonable to say that polish manufacturers don't want us home detailers to go willy-nilly applying their researched and developed to death polish onto our cars using a machine polisher. They're afraid we will bring Armageddon down on our car, burn through the clear coat, drop the polisher on the paint, or have the polisher go catastrophically skidding and hopping across the bodywork like some cursed, demented Easter bunny!

So, they strongly recommend we 'amateurs' apply their polish to our cars, only by hand, as they feel we might manage not to destroy our paintwork in doing so. That said, I'll be honest and admit I find hand polishing more therapeutic at times. You don't have that tiresome droning noise of the machine polisher ringing through your earholes or trying to vibrate your hands off at the wrists for a few hours!

The key advice with using a machine polisher is just to be careful. If you're smart and a little bit brave, bringing along that large dose of common sense and 'soft hands,' you absolutely can safely buff up your ride with a budget machine polisher if you fancy having a crack at it.

> ***Top Tip*** To maximise the chance of achieving reliable results, do some machine polish practice on a sacrificial (uncherished) vehicle using a fine or medium polish liquid before you let loose on your pride and joy. After a minute,

buff off the polish residue and check the results. If you'd like more shine, just apply a little more pressure on the next practice pass, but never press down too hard. It's much easier to work your way to the desired finish by incremental steps than having to repaint if you initially press down too hard.

I practised on the roof of my VW Golf daily run-around. I focused on keeping the polishing bonnet flat on the surface of the roof to avoid splatter. I also started the polisher approximately 2cm (1in) in from a panel gap or trim piece and worked away from it rather than toward it. To my surprise, I experienced zero demented Easter bunny activity, and my clear coat escaped fully intact!

To Compound & Ceramic Coat or Not

Now, I get that not everyone has the time or the inclination to perform a complete DIY 20-Stage Paint Correction Procedure, including a cutting compound stage and finishing with a double ceramic coating. The decision will depend on the condition of your paintwork and your time restraints.

Fortunately, there's a cracking, quicker alternative that comes pretty darn close to matching the gloss and shine of a 20-stage process, if not the level of protection. The two options for paint correction I cover in this chapter take you through one process requiring no cutting compound or ceramic coating stages. A second process includes a cutting compound and ceramic coating stages - for ease of reference, let's call them Paint Protection '*Soft*' and Paint Protection '*Hard.*' I've done both '*Soft*' and '*Hard*' processes on my Porsche Cayman, and I'm happy to report they both WORK gloriously.

You've come this far, and maybe you're leaning toward stick or twist in your decision making. So, to help you tip the balance one way or the other, here are a few words on what cutting compound and ceramic coating will bring to the table.

What Is Compound & Ceramic Sealant?

Cutting Compound: *'Meguiar's Ultimate Compound':*

Put simply, a cutting compound is an abrasive material suspended in a paste, used to restore car paintwork. The advantage with this is that it's clear coat safe and cuts as fast as harsh abrasives without scratching or hazing. It authentically restores colour clarity to faded and neglected finishes.

I used it as a 2-stage compound/polish treatment to my tired-looking Cayman front bumper. The results were so pleasing that I abandoned my plan to have the bumper expensively repainted by a body shop. You can use either a hand applicator or an electric polisher with it. The flip-top cap makes it easy to control how much liquid solution you dispense, and it's great for removing stubborn 'bird bomb' residue.

Ceramic Coating: *'Hybrid Solutions Ceramic Spray'*:

Essentially, this is science in a spray bottle! Ceramic coatings were the preserve of the pro detailing industry, and the relatively high price of the products used to be an obstacle to home detailers. However, all that has changed, ceramic coating technology has advanced leaps and bounds and is now accessible to all. This coating really will help your car stay clean for longer; it's formulated with SiO_2 and delivers phenomenal hydrophobicity plus 12 months of ceramic protection when used correctly. The synthetic wax polymers will increase your car's depth of colour and gloss and give a brilliant mirror-like shine, and the exquisite fruity fragrance is a treat on the nostrils!

Top Tip I've found this works best in overcast, shady conditions when your paintwork is cool to the touch. The beauty of it is that you can use it on all exterior surfaces except convertible tops, though I don't use it on my wheels.

Decision Made

So, now you've decided your paintwork needs either the Paint Protection *'Soft'* or Paint Protection *'Hard'* treatment. Brilliant! Let's get cracking on the next steps.

The Sun

Sorry to put a downer on the sun, but I need to refer you back to Chapter 1, *'Sunshine – the Safe Wash Enemy.'* Never start or finish your Safe Wash or Paint

Correction Processes in direct sunlight. Pick a couple of days when the forecast is dry and mild, preferably overcast when your paintwork is cool to the touch.

Paint Correction & Protection Kit List*

Detailing Hardware Required:

- Carry Caddy
- Nitrile Gloves
- Kneeling Mats
- Folding Stool
- Pressure Washer Kit & Extension Lead or Hosepipe and Nozzle
- Wash and Rinse Buckets & Grit Guards
- Wash Mitt
- Wheel Wash Bucket & Kit
- Car Blower
- Detail Guardz
- Handheld Foam Sprayer (if not using a Pressure Washer)
- Exterior Detailing Brush
- Microfibre Cloths (a bag full)
- Extra-Large Drying Towel
- Plush Buffing Towel
- Plastic Jug
- Window Squeegee
- Auto Masking Tape
- Clay Mitt
- Machine Polisher
- Wool Polishing Bonnets x 4 **or** Microfibre Polish Applicator Pads x 12
- Trim Removal Tools
- Earplugs
- Soft Toothbrush
- Eye Protection
- Face Mask
- Old Bath Towel

Detailing Liquids Needed:

- Rust Repellent
- Autoglym Pure Shampoo
- Snow Foam
- Perl Dressing Spray
- All-Purpose Cleaner (APC) Spray (diluted 1:5)
- Isopropyl Alcohol (IPA) Spray (diluted 70:30)
- WD-40 Spray
- Fallout Remover Spray
- RUPES UNO PROTECT One Step or Cutting Compound
- Autoglym Super Resin Polish
- Ceramic Spray Coating
- Tar & Glue Remover Spray
- Microfibre Detergent

*The above kit list is relevant for both Paint Protection 'Soft' and Paint Protection 'Hard' processes; pick out the kit relevant to your project.

A Word on 'Pure' Shampoo

You may have noticed that the shampoo in the above kit list has changed from Chapter 1. It's because 'Pure' will dependably strip previously applied layers of wax and polish. You only want to do this in preparation for paint correction and protection. 'Pure' is a safe, easy-to-use, high foaming shampoo that cleans thoroughly without streaking. It contains no wax or gloss enhancers, which leaves nothing behind - ideal for paint correction prep.

Preparation

Remember to get those beers in the fridge ready for when you finish and grab a bottle of water or Thermocafé mug of coffee to sustain you.

Remove your rings and watch, and make sure you aren't wearing anything zippy or with buttons or poppers that could scratch your car while working up close to it. Position your car, preferably with room to safely work around it. Place your Detail Guardz under the outer edges of the tyres. Get your Paint Correction & Protection Kit assembled, load your Carry Caddy with your Spray Potions and Microfibre Cloths, and have your Caddy handy.

Breaking Up the Work

Paint Correction and Protection is a commitment as there's a fair bit of work to get through. I find it easiest to split the tasks up over 2-3 days to focus on getting each step right without rushing and making a pig's ear instead of a silk blouse! You'll only need a third day if you're ceramic coating your ride - the reason being you need to allow the first coat of ceramic spray sealant to cure for 24 hours before applying a second coat the following day. So, I do:

- Day 1: The Safe Wash steps 1-15 (see Chapter 1)
- Day 2-3: Paint Correction and Protection steps

Day 1

Safe Wash

Focus on steps 1-15 Maintenance Safe Wash as explained in Chapter 1. Miss out Step 11 – waxing, as you don't need to wax the car before paint correction! If possible, move the car undercover at the end of Step 15. If that's not possible, not a problem, you can sort the car out at the beginning of day 2.

Part 2 - Paint Correction *'SOFT'*

Day 2: Paint Correction *'Soft'* or *'Hard'*

Now we come into new territory for your Driveway Warrior exploits! Don't worry; you've *GOT* this! Retrieve all your Kit and your Carry Caddy and check that your Detail Guardz are in position under the outer edge of your tyres.

I like to split the day 2 processes into two stages:

- Stage 1: Decontamination (Steps 16 – 20). This stage applies whether you've elected for Paint Protection *'Soft'* or Paint Protection *'Hard.'*
- Stage 2: Paint Correction and Protection (there are different versions of Step 21 for *hand* and *machine* application).

Stage 1: Decontamination Steps 16-20

Step 16 - Car Blower: Call me fussy, but I like to start by blowing off any dust that has settled on the car. Grab your Car Blower, trail the extension lead over your shoulder so that it doesn't drag scratchily over your paintwork, and go around the car air blasting off any pesky dust from the surface. Don't let the blower nozzle hit your car. Remember to smile politely at passers-by who may gaze in wonder as you frantically blow away at a seemingly bone dry and gleaming vehicle! Move the Blower and extension lead away from the car out of your way when done.

Chemical Decontamination:

Step 17 - Fallout Remover: With your Gloves on, seize your Fallout Remover Spray. This stuff is formulated to attack and dissolve contamination particles that have bonded to your paintwork, which safe wash will not remove. Word of warning: use a face mask, as all Fallout Removers reek of sulphur! Just hold your nose, then lightly and evenly mist down all your paintwork. Work panel by panel methodically, top to bottom, focusing more of the spray on the lower third of the car where fallout particles are heavier. Allow it to soak for 2 minutes, but don't let it dry on the paint.

Observe the immensely satisfying sight of the purple chemical reaction as those peeving particles start to dissolve. Then simply rinse thoroughly top to bottom with your Pressure Washer wide fan nozzle or Hosepipe Nozzle. If you haven't passed out yet, on we go!

Next, I like to test a small paint area to check how the Fallout Remover has performed. Grab your drying towel and dry off the driver's side door only. Run your un-gloved fingertips lightly over the door, particularly the lower third area and check if it feels any smoother. If the whole area now feels nice and smooth, you can miss out Step 19 - Manual Decontamination, but chances are it'll still feel slightly rough under your fingertips, in which case you will need Step 19. When done, move your Fallout Remover and Pressure Washer or Hose Nozzle out of your way.

Step 18 - Tar & Glue Remover: If your paint is suffering from tar deposits on the lower third of the bodywork, you'll need to treat it with Tar Remover, already decanted into a spray bottle. Lightly and evenly mist down the lower third of your paintwork, where tar deposits are concentrated. Work panel by panel methodically, allowing it to soak for 2 minutes, but don't let it dry on the paint. Then once again, rinse thoroughly with your Pressure Washer wide fan nozzle or Hosepipe Nozzle. When done, move your Tar Remover and Pressure Washer or Hose Nozzle out of your way.

Manual Decontamination

Step 19 - Clay Mitt: Now we're getting into serious detailing territory! You arrive at Step 19 because your paintwork still feels a little too rough under your fingertips for your liking. Fortunately, you don't need to struggle with those pesky, pricey little lumps of car clay that go rock hard in the cold and have to be binned the first time they're dropped. Grab your clean Wash Bucket and Grit Guard, add a capful of Pure Shampoo, and half-fill the bucket with warm water, swirling the water to mix.

Seize your Clay Mitt, dunk it in the soapy water to lubricate thoroughly, and gently work the Clay Mitt over the paintwork using light pressure only. Work panel by panel methodically from top to bottom in straight lines only, *keeping* the

Clay Mitt super-well-lubricated with soapy water (a dry Clay Mitt will scratch our paint). Used in this way, the Mitt will help to safely remove bonded contaminants, ingrained dirt and tar spots from your paint.

It's simple, effective, five times faster and lasts five times longer than a clay bar. If you drop the Mitt, give it a thorough rinse in the wash bucket and carry on until you've covered every inch of the paintwork. Next, take your Pressure Washer or Hose Nozzle and rinse the car down thoroughly from top to bottom. Finish this step by blasting the excess rinse water off with your Car Blower, then take your Extra-Large Drying Towel and dry the whole car down, ready for Step 20. Finally, move your Pressure Washer or Hosepipe, Wash Bucket, Clay Mitt and Car Blower away from the car out of your way.

Final Paint Cleanser

Step 20 - IPA Wipe Down: This is the final cleaning step before you get to break out your cutting compound and or polish. Take your IPA Spray bottle, diluted from the 99.9% stuff, down to a 70:30 mix (IPA to De-ionised Water), giving you the ideal 70% proof IPA panel wipe. Also, grab a couple of clean, different coloured Microfibre Cloths. Then, working methodically from top to bottom, lightly and evenly spritz the IPA over the paint one panel at a time. Spread the IPA over the paint with one Cloth and wipe it dry with the other, folding your drying Cloth as you go. The IPA leaves no residue and will flash away easily with the help of your drying Cloth, and it will remove any remaining detergents or chemicals lingering on your paintwork. Your paint is now as clean and bare as possible, giving you an ideal base to lay down the Compound or Polish. Move your IPA Spray and used Microfibre Cloths away from the car out of your way.

Congratulations, you are done with Stage 1: Decontamination (steps 15 – 20)!

Get ready to move onto Stage 2: Paint Correction and Protection.

Stage 2: Paint Correction and Protection (*there are different versions of Step 21*)

Okay, let's first get into our so-called Paint Protection '*Soft.*' It's the quicker of our two Paint Correction and Protection options; you'll still need to put work in, though it will reward you mightily! A quick recap: *RUPES UNO PROTECT One Step Polish & Sealant* is a true 'all-in-one' compound, polish and protectant product that removes light defects and produces a breath-taking high gloss mirror finish on your paintwork in one step! It uses a rich blend of polymer, silicone, and carnauba to put down a durable, protective layer capable of lasting up to three months. Pretty handy, right?

Next up, I'll go over the method for application with a hand applicator, followed by the method for application with a budget machine polisher, and you can decide which way you want to go. As indicated earlier, the clear advantage of the machine polisher is the speed with which it gets the job done, taking about half the time of the hand applicator.

Heads Up

At this stage, I need to give you a *heads up* that the following pages do include some repetition of the necessary processes using the *'Soft'* and *'Hard'* methods. I incorporate techniques for application by *'hand'* and *'machine polisher,'* as understandably, some readers may prefer to skip straight to their preferred *'Soft'* or *'Hard'* method, then skip again to their preferred *'hand'* or *'machine polisher'* system.

To Tape Up or Not

This is a tough one: deciding whether to tape off plastic or rubber trim or panel gaps before you polish can be a head-scratcher. There are a few reasons for masking off parts of the car, including minimising contamination of the plastic or rubber trims by touching them with your applicator pad or machine pad, which you will then have to faff around cleaning off. Also, to reduce the risk of damaging plastics or trim, it's possible to scuff or discolour delicate trim, particularly when using a machine polisher.

Another consideration is specialist auto detailing tape is pretty pricey, in my view, and it comes down to a personal preference and, to a great extent, your polishing technique. A guiding principle would be that if you can avoid touching it with your applicator pad or machine pad, there's no need to cover it in detailing tape.

So, that said, I tend not to mask off anything when using a hand applicator pad, and I keep away from trim and panel gaps, but I do mask off some parts when I use a machine polisher.

I'll cover this when we come to the machine polishing section at Step 21B.

Step 21A - *Hand Applicator Method*

Paint Correction and Protection '*Soft*' by hand: RUPES UNO PROTECT One Step Polish & Sealant

Snag your Caddy. Load it with half a dozen or so Microfibre Applicator Pads, RUPES UNO PROTECT One Step, Soft Buffing Towel, Exterior Detailing Brush, Flat-ended Trim Removal Tool, IPA Spray (70:30) and a few clean Microfibre Cloths. Put your Kneeling Mats and Stool in position.

Lay Down & Spread the Polish

Wearing your Gloves, shake the UNO well and apply two pea-sized dots to your Hand Applicator Pad. Pick your starting point 2cm (1in) away from a panel gap. Working methodically from top to bottom, panel by panel, work the UNO onto the paintwork with moderate pressure, using a circular motion, for no more than one minute. Do this in sections approximately six times the size of your applicator pad in an overlapping crosshatch pattern.

I find that by restricting the size of my working area in this way, it's easier to keep the Pad away from plastics and trim parts or panel gaps. I start approximately 2cm (1in) in from a panel gap or trim piece and work away from it rather than toward it. Then when you approach another panel gap or trim piece, use the same method of starting just inside the gap and work away from it, meeting your applied polish in the middle.

Polish 'Set Up' & Buffing

Allow a minute for the UNO to 'set up' before buffing off the worked section with your dedicated Soft Buffing Towel, folding the Towel frequently as you go. The towel folding is important because it acts as a padded barrier, reducing the pressure of your hand pressing down on the surface. I find this buffing stage incredibly satisfying, as you start to see the fruits of your labour, and in my experience, the UNO does not 'dust up' when used correctly. Go around the whole car in this way, changing your Applicator Pad every three or four panels to avoid it becoming polish-clogged.

Oops, a Bit Too Much Polish!

You'll know if you've put too much Polish on the paint because your Buffing Towel will start to drag annoyingly. If you do get in a pickle with a little too much Polish on the paint, don't panic. Spritz the overloaded section with IPA and wipe it down with a clean, dry Microfibre Cloth, then slightly reduce the amount of Polish used on the next section and carry on. If you accidentally get Polish in a panel gap, straightaway take your Detailing Brush, work it into the gap, and it should remove the polish residue, provided it hasn't gone too hard. If it has gone a bit hard, grab your flat-ended Trim Removal Tool, fold a clean Microfibre Cloth over the flat end and ease out the irksome residue.

Hand Applicator *'Soft'* Process Done

You are DONE; congratulations! By the time you've finished the whole car, your biceps will be bulging impressively; your ride will look resplendent, and it will be protected for three whole months. Great job, you! RUPES recommend you allow the UNO to cure and don't get the paintwork wet for four hours after use. Why would you anyway, right? Oh Lord, please don't let it rain.

Photos

You *so* need to capture some sexy shots of your masterpiece at this point. Have a quick peek back at the end of Chapter 1, where I offer hints and tips for achieving wicked driveway snaps!

A final comment on RUPES UNO PROTECT One Step: I've found it excellent at reinvigorating previously applied ceramic coatings. Let's say you ceramic coated your ride a year or two ago, and the paintwork now needs refreshing and bringing back to life. RUPES UNO will do that superbly for you in one step while adding that three months of protection.

Step 21B - *Budget Machine Polisher Method*

Paint Correction Protection *'Soft'* by machine: RUPES UNO PROTECT One Step Polish & Sealant

Masking off: As indicated previously, we need to do some masking off before breaking out the machine polisher. I tend to go for the sky-blue variety of 3M auto masking tape; it's comparatively reasonably priced and widely available in different widths. Its low-tack rubber-derived adhesive doesn't lift even when wet or leave behind any messy residue. I use the 19mm stuff as it's easy to curve it around the contours of a headlight or an emblem, and it's wide enough to cover a strip of rubber trim underneath a window, for instance.

It's useful to remember that the trim and rubbers found on most modern cars are far more durable than those found on older or classic cars. So, if you're working on a classic car, you'll need to spend more time on the masking. I never use more than one roll of tape when masking, as I only want to cover the key areas, and masking off isn't an art project. So I try and limit my masking off time to 15 minutes. As I suggested on Page 134, a guiding principle would be that if you can avoid touching it with your machine pad, there's no need to cover it in detailing tape. It is down to personal preference what you think needs masking on your car – if anything. Engage the old common sense and give whatever tape you use a firm press with your thumb tip to secure it in place.

Gather the Machine Polishing Kit: Grab your Machine Polisher, Extension Lead and Wool Polishing Bonnets x 4. Load your Caddy with two Microfibre Applicator Pads, RUPES UNO PROTECT One Step, Soft Buffing Towels, Exterior Detailing Brush, Flat-ended Trim Removal Tool, IPA Spray (70:30) and several Microfibre Cloths.

You'll also need your Auto Masking Tape, Car Blower, Soft Toothbrush, Eye Protection, Ear Plugs and an Old Bath Towel.

Put your Kneeling Mats and Stool in position, and place your Detail Guardz under the outer edges of your tyres so that the extension cable won't snag and slow you down. Do your masking off; remember that masking isn't an art project, but spending some time on this task will serve you well. Securely fix a Polishing Bonnet to the Polisher, ensuring it's nice and snug.

Top Tips You can make the Polisher more comfortable to use by wrapping bubble wrap around the handles, secured with tape, to reduce vibrations through your hands, as seen in the picture on page 129. As a precaution against Polish spatter, drape the old clean Bath Towel across your windscreen, securing it behind the wipers.

Stuff your Earplugs in to protect your hearing from the machine drone. Gloves and Eye Protection on, shake the UNO well and apply three pea-sized dots evenly across your Wool Polishing Bonnet. Hook the extension lead over your shoulder so it doesn't trail scratchily over the paintwork.

Only start or stop the machine directly on the paint to avoid any polish sling. Pick your starting point just inside a panel gap and place the Polishing Bonnet on the paintwork and switch it on while holding the machine firmly. Keep the Polishing Bonnet flat to the surface to avoid spatter. NEVER lift the machine away from the paint while it's switched on, or it WILL make a mess, which will slow you down while you have to clean up the spatter.

Lay Down & Spread the Polish

Work from top to bottom, panel by panel methodically, using light pressure. Work the UNO onto the paintwork in sections approximately six times the Polishing Bonnet's size. Initially, spread the Polish evenly within your working zone, then work slowly across the working zone, doing three passes in an overlapping crosshatch pattern. Spend no more than one-minute machine polishing on any one working zone.

I start approximately 2cm (1in) in from a panel gap or trim piece and work away from it rather than toward it. Then, when you approach another panel gap or trim piece, use the same method of starting just inside the gap and work away from it, meeting your previously applied Polish in the middle. I find that by restricting the size of my working area in this way, it's easier to keep the Polishing Bonnet away from plastics and trim parts or panel gaps.

Top Tip Don't try to Machine Polish into tight areas; take your Microfiber Applicator Pad and shape it around your finger(s) to work the Polish into those tight spots.

Polish 'Set Up' & Buffing

Allow a minute for the UNO to 'set up' before buffing off the worked section with your dedicated Soft Buffing Towel, folding the towel frequently as you go. The towel folding is important because it acts as a padded barrier, reducing the pressure of your hand pressing down on the surface. I find this buffing stage incredibly satisfying, as you start to see the fruits of your labour, and in my experience, the UNO does not 'dust up' when used in this way.

When you've finished Soft Towel Buffing one working section, reload your Polishing Bonnet with three more pea-sized blobs of UNO and repeat the process on the next section, approximately six times the size of the bonnet. I find it helps to change the Bonnet after approximately 25% of the car has been polished; the Wool Bonnets seem to work best when not polish-clogged. Go around the whole car in this way until you've polished every bit.

Top Tip If you run short of Wool Polishing Bonnets due to them becoming polish-clogged, you can clean one while it's still on the machine to enable you to carry on. BUT you must clean it well away from the car, and I mean literally on the other side of a fence, gate, or house! Take your Car Blower, Soft Toothbrush and put your Eye Protection on. Place the Machine Polisher on a firm surface with the Bonnet facing up. Switch on the Polisher and run the Toothbrush head over the spinning Bonnet, slowly running the brush from the centre to the outer edge of the Bonnet. Depending on how polish-clogged it is, the Bonnet may furiously sling Polish everywhere, including all over you, so make sure you're not wearing your designer gear!

Finish by air blasting the spinning Bonnet with your Car Blower, moving the torrent of air from centre to outer edge for about ten seconds. Hey presto, you have one Polishing Bonnet good to go again!

Oops, a Bit Too Much Polish!

You'll know if you've put too much polish on the paint because your buffing towel will start to drag annoyingly. If you get in a pickle with too much polish on the paint, follow the same steps for 'Oops, a Bit Too Much Polish' on Page 136.

Photos

You are DONE; congratulations! You *so* need to capture some sexy shots of your masterpiece at this point. Have a quick peek back at the end of Chapter 1, where I offer hints and tips for achieving wicked driveway snaps!

Part 3 - Paint Correction *'HARD'*

Step 21C - *Hand Applicator Method*

Paint Correction Protection *'Hard'* by Hand: 4 Stage Process

Okay, so we arrive at *Paint Protection 'Hard'* because you're not happy with the condition of your paintwork. You've decided it needs serious attention, and you've resolved to put the time and effort in to get it sorted yourself. Let's do this Driveway Warrior!

Grab your Caddy and load it up. You'll need six Microfibre Applicator Pads, Cutting Compound, Super Resin Polish, Soft Buffing Towels x 2, Exterior Detailing Brush, Flat-ended Trim Removal Tool, IPA Spray (70:30), several Microfibre Cloths and Ceramic Spray Coating.

Put your Kneeling Mats and Stool in position.

Stage 1 - Cutting Compound by Hand

With your Gloves on, shake the *Meguiar's Ultimate Compound* well and apply two pea-sized dots to your Applicator Pad. Pick your starting point just inside a panel gap and work methodically away from the panel gap from top to bottom, panel by panel. Work the compound onto the paintwork with moderate pressure using a circular motion. Do this in sections approximately six times the size of your Applicator Pad, in an overlapping crosshatch pattern for one to two minutes. Be careful to apply even pressure over the Pad to avoid pressing down with your fingertips in a concentrated area.

I find that by restricting the size of my working area in this way, it's easier to keep the Pad away from plastics and trim parts or panel gaps. See the rationale for this in Part 2, Page 135, 'Lay Down & Spread the Polish.'

Before the compound dries, buff off your worked section with your dedicated Soft Buffing Towel, folding the towel frequently as you go. I find this buffing stage incredibly satisfying, as you start to see the fruits of your labour, and in my experience, the Meguiar's Compound happily does not 'dust up' when used correctly. Go around the whole car in this way, changing your applicator pad every three or four panels to avoid it becoming compound clogged.

A Little Too Much Cutting Compound

You'll know if you've put too much Compound on the paint because your Buffing Towel will start to drag annoyingly. If you do get in a pickle with a little too much Compound on the paint, don't panic. Spritz the overloaded section with IPA and wipe it down with a clean, dry Microfibre Cloth and carry on. If you do accidentally get compound in a panel gap, straightaway take your Detailing Brush, work it over the gap, and it should remove the compound residue, provided it hasn't gone hard. If it has gone hard, grab your flat-ended Trim Removal Tool, fold a clean Microfibre Cloth over the flat end and ease out the irksome residue. Move your Compound, Applicator Pads, compound Buffing Towel and Cloths away from the car out of the way.

Stage 2 - Polishing by Hand

Here, the method is identical to 'Cutting Compound by Hand' above. Shake the *Super Resin Polish* well and apply two pea-sized dots to a fresh Applicator Pad. Pick your starting point just inside a panel gap and work methodically away from the panel gap from top to bottom, panel by panel. Work the Polish in a thin layer onto the paintwork with moderate pressure using a circular motion. Do this in sections approximately six times the size of your Applicator Pad, in an overlapping crosshatch pattern for one to two minutes.

Allow it to dry for a couple of minutes before buffing off your worked section with a fresh dedicated Soft Buffing Towel, folding the Towel frequently as you go. Go around the whole car in this way, changing your Applicator Pad every three or four panels to avoid it becoming polish-clogged. In my experience, Super Resin Polish happily does not 'dust up,' provided you don't apply too much.

A Drop Too Much Polish

You'll know if you've put too much Polish on the paintwork because your Buffing Towel will start to drag annoyingly, and the polish may start to 'dust up.' If you do get in a pickle with a little too much polish on the paint, don't panic. Follow the same steps explained in 'A Little Too Much Cutting Compound' on Page 141.

Stage 3 - IPA Wipe Down

You might think this step is overkill, but I suppose that's the nature of detailing, and this is the final step before you break out your ceramic coating. A final IPA wipe down will remove any chemicals or oils lingering on your paintwork from the compounding and polishing stages, leaving the surface as clean and bare as possible. This will give your ceramic coating the best possible chance to bond to your paintwork and do its magic of achieving 12 months of protection. Follow the same steps in Part 2 on Page 133, 'Step 20 - IPA Wipe Down.'

Stage 4 - Ceramic Paint Sealant 12 Months Protection

Finally, we arrive at the last stage; you're almost home and dry. For this, you'll need your Ceramic Spray, a clean Microfibre Cloth, Soft Buffing Towel and IPA

Spray. Remember that the *Turtle Wax Hybrid Solutions Ceramic Spray Coating* works best in overcast, shady conditions when your paintwork is cool to the touch. I would not attempt to apply it in sunny conditions when your paintwork is hot.

Ceramic Sealant First Coat

Trial and error have taught me that less is more with this stuff. My favoured method is to shake the bottle well and lightly spritz a clean, folded Microfibre Towel with Ceramic Spray. Spread the ceramic coating over the panel using the Cloth in this way, working methodically, one panel at a time, then buff off the panel with your dedicated folded Soft Buffing Towel. As you move to a new panel, refresh your Microfibre Cloth with more Ceramic Spray and carry on until you've done the whole car, folding your Soft Buffing Towel and buffing as you go.

The beauty of this Ceramic Spray is that you can use it on all exterior surfaces except convertible tops. However, I don't use it on my rims (there is a dedicated ceramic coating for wheels which I cover in Chapter 6). You'll know if you apply too much ceramic spray to a panel because it will start to smear and become a pain to buff off. If it does smear up, no problem; spritz some IPA over the smeared area, wipe off with a clean Microfibre Cloth and carry on. Spritz slightly less ceramic spray on your Microfibre Cloth for the next section – sorted!

Ceramic Sealant Second Coat - 24 Hours Later

Important to note that you need to allow the first coat to cure 24 hours before applying a second coat or washing the car. By applying a second coat, you maximise your chance of achieving the full 12 months of protection. Unless you're lucky enough to be able to store your car in a garage overnight, what I like to do after the 24 hours, is take my car blower and go around the car, giving it a quick air blast to remove any dust that's settled on it. You'll find you can do the second coat faster because the first coat enables your buffing towel to slip effortlessly over the paintwork. Plus, your technique will have refined, and your speed will increase from doing the first coat.

Photo Time

You are DONE; congratulations! You *so* need to capture some sexy shots of your masterpiece at this point. Have a quick peek back at the end of Chapter 1, where I offer hints and tips for achieving wicked driveway snaps!

Step 21D - *Budget Machine Polisher Method*

Paint Correction Protection '*HARD*' by Machine: (Cutting Compound, Polish & Ceramic Coating) 4 Stage Process

Grab your Machine Polisher, Extension Lead, Car Blower and Wool Polishing Bonnets x 4. Load your Caddy with two Microfibre Applicator Pads, Compound, Super Resin Polish, Ceramic Spray, Soft Buffing Towel, Exterior Detailing Brush, Flat-ended Trim Removal Tool, IPA Spray (70:30), Microfibre Cloths, Auto Masking Tape, Toothbrush and Bath Towel. Retrieve your PPE - Eye Protection and Ear Plugs.

Put your Kneeling Mats and Stool in position. Do your masking off; remember that masking isn't an art project, but spending some time on this task will serve you well. Place your Detail Guardz under the outer edges of your tyres so that the extension cable won't snag and slow you down. Securely fix a polish Bonnet to the Polisher; make sure it's nice and snug.

Top Tips You can make the Polisher more comfortable to use and protect the windscreen from polish spatter – see Top Tips on Page 129.

Stage 1 - Cutting Compound by Machine

Stuff your Earplugs in to protect your hearing from the drone of the machine. Gloves and Eye Protection on, shake well the Compound Solution and apply three pea-sized dots evenly across your Wool Bonnet. Hook the extension cable over your shoulder so it doesn't trail scratchily over the paintwork. Pick your starting point just inside a panel gap, place the Bonnet on the paintwork, and switch it on while holding the machine firmly. Only ever start or stop the machine directly on the paint to avoid any polish sling. Keep the Bonnet flat to the surface to avoid spatter. NEVER lift the machine away from the paint while it's still switched on, or it WILL make a mess, which will slow you down while you have to clean up the spatter.

Work methodically from top to bottom, panel by panel, using light pressure, and work the Compound onto the paintwork in sections approximately six times the Bonnet's size. Initially, spread the Compound evenly within your working zone, then work slowly across the working zone, doing three passes in an overlapping crosshatch pattern.

Spend no more than one minute of machine compounding on any working zone. I find that by restricting the size of my working area in this way, it's easier to keep the Bonnet away from plastics and trim parts or panel gaps. I start approximately 2cm (1in) from a panel gap or trim piece and work away from it rather than toward it. Then when you approach another panel gap or trim piece, use the same method of starting just inside the gap and work away from it, meeting your previously applied polish in the middle.

Top Tip Don't try to machine compound into tight areas. See the Top Tip on Page 138.

Before the Compound dries on your working zone, buff off your worked section with your dedicated Soft Buffing Towel, folding it frequently as you go. The towel folding is important because it acts as a padded barrier, reducing the pressure of your hand pressing down on the surface. I find this buffing stage incredibly satisfying as you start to see the fruits of your labour. In my experience, the compound does not 'dust up' when used in this way.

When you've finished Soft Towel Buffing one working section, reload your bonnet with three more pea-sized blobs of Compound. Repeat the process on the next section, approximately six times the size of the Bonnet. I find it helps to change the Bonnet after approximately 25% of the car has been compounded; the Wool Bonnets seem to work best when not compound-clogged. Go around the whole car in this way.

Top Tip If you run short of Wool Bonnets due to them becoming compound-clogged, you can clean one while it's still on the machine to enable you to carry on. See the Top Tip described in part 2 on Page 139.

A Squirt Too Much Compound

You'll know if you've put too much Compound on the paint because your Buffing Towel will start to drag annoyingly. If you get in a pickle with a little too much

Compound on the paint, don't panic. Follow the same steps explained in 'A Little Too Much Cutting Compound' on Page 141.

Stage 2 - Polishing by Machine

The method for your *Super Resin Polish* is identical to the Cutting Compound method. So, ensure your gloves, eye protection and earplugs are all deployed and follow the same steps shown in 'Stage 1 - Cutting Compound by Machine' on Page 144.

A Splodge Too Much Polish

You'll know if you've put too much Polish on the paint because your Buffing Towel will start to drag annoyingly. If you get in a pickle with a little too much Polish on the paint, don't panic. Follow the same steps explained in 'A Little Too Much Cutting Compound' on Page 142.

Stage 3 - IPA Wipe Down (Same for Hand or Machine Polishing)

"What, IPA again?" I hear you cry.

You might think this step is overkill, but I suppose that's the nature of detailing, and this is the final step before you break out your ceramic coating. Follow the steps explained in 'Step 20 - IPA Wipe Down' on Page 133.

Stage 4 - Ceramic Paint Sealant 12 Months Protection (Same for Hand or Machine Polishing)

Finally, we arrive at the last stage; you're almost home and dry. For this, you'll need your Ceramic Spray, a clean Microfibre Cloth, Soft Buffing Towel, and IPA Spray. Remember that the *Turtle Wax Hybrid Solutions Ceramic Spray Coating* works best in overcast, shady conditions when your paintwork is cool to the touch. I would not attempt to apply it in sunny conditions when your paintwork is hot.

Ceramic Coating First Coat

Trial and error have taught me that less is more with this stuff. Just follow the instructions shown in 'Ceramic Sealant First Coat' on Page 143.

Ceramic Coating Second Coat – 24 Hours Later

It is important to note that you need to allow the first coat to cure 24 hours before applying a second coat or washing the car, which is why you'll need an hour or so on the third day. By applying a second coat, you maximise your chance of achieving the full 12 months of protection. Unless you're lucky enough to be able to store your car in a garage overnight, follow the same steps shown in 'Ceramic Coating Second Coat – 24 Hours Later' on Page 143.

Photos

You are DONE; congratulations! You *so* need to capture some sexy shots of your masterpiece at this point. Have a quick peek back at the end of Chapter 1, where I offer hints and tips for achieving wicked driveway snaps.

Clear Up

Once again, to the part we hate! Clear up is a necessary evil and will ensure all your precious detailing kit stays in tip-top condition, ready to go next time you need it. Start by returning all your Liquids and Spray bottles to storage, but check that all lids and spray heads are secure and keep an eye on any dilutables that need topping up. Clean out the Compound and Polish bottle lids to prevent any liquid residue from going hard before they go into storage. If you used the Handheld Foam Sprayer, rinse it out with clean water, semi-pressurise it, and spray out some clean water to clear the feed pipe and nozzle. Give it a quick dry and store it away. Grab your Car Blower, Caddy, Mats, Stool, Detail Guardz and Squeegee and give them a quick wipe down to dry and store them away.

Discard all the bucket water – my rinse water goes in the water butt. Give your three Buckets and Grit Guards a rinsing blast with the Pressure Washer or Hose Nozzle, then put them somewhere to drain upside down. Hold back the Wheel Bucket as you'll need it in a bit.

Next, gather all used Cloths, Mitts and Towels and squeeze them out if necessary. If the Cloths are mucky, including your compound or polish-laden Cloths, Applicator Pads or Machine Polisher Bonnets, chuck them in a clean bucket filled with clean, warm water, add a small squirt of washing-up liquid, and give them a short pre-wash with your gloved hand. Then squeeze them out and pop them in the washing machine at 40°C (104°F), add the Microfibre Detergent, and stick on a 30-minute cycle with a gentle spin. The only thing I wash separately from the other cloths is my Glart Soft Buffing Towel; it's a bit more delicate, so I do it at 30°C (86°F). When you take them out of the machine, give them a good shake to re-set the fibres and hang everything on a clothes dryer to dry naturally, using pegs for the mitts.

You'll be left with your Pressure Washer, Brushes and Wheel Bucket. Dump all your Brushes in the Wheel Bucket and give them a quick rinsing blast with the Pressure Washer or Hose Nozzle. Discard that rinse water in the water butt, add a squirt of washing-up liquid to the bucket, and half-fill it with hot tap water, swirling it around to mix. Spend a few minutes with your gloved hands in the soapy water, massaging the Brushes clean. You'll find the Tyre Brush gets the muckiest, so I give this one an extra squirt of washing-up liquid and work it in. Dump the soapy water and give the Brushes a final rinsing blast in the bucket with the Pressure Washer or Hose Nozzle and discard the rinse water.

I then give all the Brushes a good wrist flick to fling off excess rinse water, and I lay them on an old Towel under the radiator to dry or in the sun if you're lucky. Give your three clean Buckets and Grit Guards a quick wipe to dry and store them away with their lids on to keep the insides clean. Give your Pressure Washer and Accessories, including the power cable and Machine Polisher if used, a wipe down and store them away. The next day, when your Towels, Cloths, Applicator Pads,

Mitts and brushes are fully dry, store them away, keeping your Wheel Brushes in your Wheel Wash Bucket.

Reflection

So, you've successfully completed a mammoth, meticulous paint correction/restoration and protection detailing project, probably the most labour intensive but rewarding of all detailing activities you'll ever undertake. Super *well done*; you are AMAZING! You should be bursting with pride at the astonishing transformation of your paintwork; you've banished that dull, flat, oxidised paint finish forever. People may assume you've had your car re-sprayed, and that is the ultimate compliment! You fully deserve to snag those icy beers from the fridge and indulge in an extra-long feet-up rest and relaxation as you happily reflect on the *whopping* **£1,231 ($1,673)** saving on what the pro detailer quoted for this work. This is the average of three quotes from UK pro detailers - how cool is that?

As the days go by and you bask in contentment at the gorgeously detailed ride you've created and are enjoying to the maximum, your detailing instincts may spur you on to the next projects. Read on in Chapter 9 to find out what happened when my Cayman's headlights *disastrously* let me down in the pitch blackness of rural Surrey country lanes at midnight in *Handsome Headlight Restoration* ... I know, you can't wait!

BTW, remember my VW Golf I mentioned at the beginning of the chapter where the cheeky used car buyer offered me £900 ($1,218)? Well, he spurred me into action; I taught myself how to do the paint correction and restoration, then I sold the magnificently detailed car privately for £5,100 ($6,904) – happy detailing days!

CHAPTER 9

Handsome Headlight Semi-Permanent Restoration and Protection

Part 1: 7 Stages to Handsome Headlight *'OFF'* the Car Restoration

It's the *early hours*, my wife and I are tired but happily homebound to Sussex in the Cayman from a great party in rural Surrey. These deserted country roads were *made* for the Cayman and my carefully researched, award-winning high performance 'Bosch Gigalight' headlight bulbs will make short work of the oppressive pitch-black darkness of these country lanes … or so I thought!

Two miles into the remote Surrey countryside, I said to Mrs Driveway Warrior, "Babe, either my eyes or my headlights are failing because I can't see a flipping thing!"

We made a quick stop by a meadow gate, and I hopped out to investigate the source of my night blindness. Sure enough, both headlights are somewhat tarnished and cloudy, preventing even the powerful bulbs from lighting the way and making them dangerously dim.

"Oh, you pair of @$*!%#!" I breathed.

I took the next 20 miles at *'Driving Miss Daisy'* pace to make sure we arrived home safely in one piece – at least it wasn't my eyes that failed: every cloud, right? There would be no more night driving for this Driveway Warrior until the headlights mess is sorted.

Cause of Cursed Cloudy Headlights

The next day, I was straight on the case. A quick phone call to Golding Barn Garage, my friendly local Porsche indie, was most revealing. Dirt and oxidisation cause tarnishing and milky, cloudy hazing of the headlight lenses, which isn't uncommon for a Porsche approaching its twelfth anniversary. The cause is years of *rapid* driving, causing wear and tear by minor stone chips on the lens, and *prolonged* UV exposure over the same period damaging the lens.

But Do the Headlights Need Replacing?

You'd be forgiven for thinking the only way to return the headlights to their original condition would be to fork out for new jobbies.

Well, at £500 per corner for new Porsche items, *no* thank you, Signor. Surely there must be a reliable Driveway Detailing Warrior style solution, mustn't there? Oh yes, my friends, there most certainly *is*. You may have heard about headlight restoration using items like toothpaste or WD-40. Please believe me when I say these are only *temporary* fixes, and the cursed cloudy hazing will return to the lens over time.

Off-The-Shelf Kits: Pros & Cons

You *can* buy headlight restoration kits for around £30 ($40); I call these kits the *'conventional DIY route.'* These kits will work well – up to a point. What I mean by that is, in my opinion, these kits tend to play it safe by typically offering sanding options from *800 grit* up to around *3000 grit*. For those of you that aren't sandpaper aficionados, the *lower* the grit rating, the *coarser* or more *abrasive* the sandpaper is. If your headlights are in a poor state, with deep scratches and stone chip damage – 800 grit is *too* fine a grade to get the job done effectively. You're going to be at it forever trying to achieve a decent finish. So, in my view, the off-the-shelf kits work best on moderate or low levels of headlight damage.

My rationale for concluding this stems from my experience in successfully DIY refurbishing multiple sets of alloys wheels. I also prepped multiple tail lights for spray tinting. I did this at home – involving *relentless* rounds of sanding with gradually increasing grit levels, starting with the most aggressive and progressively working with finer grits to achieve as close to a perfect finish as possible.

I get why kit manufacturers play it safe with the grit grades. In my humble opinion, they do it to strike a balance between sandpaper grits that will achieve a level of finish, as opposed to more abrasive sandpaper, which the DIY first-time user may inadvertently over sand with and damage some part of the headlight.

Unconventional DIY – The Route to Success

Anyway, enough on off-the-shelf kits; let's focus on sorting *your* troublesome headlights. If you have serious cosmetic imperfections, you absolutely can tackle them and restore the lens to crystal clear condition. You will use products, some of which you may already have in your home, and your detailing arsenal - together with your usual healthy dose of Driveway Warrior elbow grease! I call this the '*unconventional DIY route,*' and the method we will use will be a semi-permanent, maintenance-free restoration.

Let's face it, oxidised, cloudy, pitted headlights are not only dangerous, but they also *look* terrible. They will drag down the appearance and value of your car no matter how beautifully detailed it may be, and that's not on, right? So, you know what to do, Driveway Warrior; let's get cracking on making them *dazzle* ...

Climate Control

Weather, weather, always with the *blinking* weather! Most of the advice given in previous chapters about detailing in ideal weather conditions applies to your headlight restoration project, with *some* exceptions. May I refer you to Chapter 1, '*Sunshine – the Safe Wash Enemy.*' Never start or finish your Detailing Processes in direct sunlight if you want to avoid issues with your detailing liquids drying out too quickly on your car and screwing up the quality of your finish.

Pick a day when the forecast is dry and warm with no breeze. We want to give ourselves every opportunity to achieve near-perfect results, right? However, if you're working with the headlights *on* the car outside, you need to do this on a *warmish*, not hot, day with *no* breeze. This is to give the Clear Coat spray the best chance to bond, and as for trying to spray paint in a breeze – please don't go there!

Also, *don't* make the same mistake I made by working with the protective Polythene Dust Sheet in the wind. It becomes a total nightmare to lay the sheet down and will slow you down and frustrate you no end – who needs that if you can avoid it?

Handsome Headlight Restoration Kit List

We do this project by hand – no drill or polishing machine required!

Detailing Hardware Required:

- Carry Caddy
- Nitrile Gloves
- Microfibre Cloths (a bag full of clean cloths)
- Exterior Detailing Brush
- Car Blower & Extension Cable
- Kneeling Mats
- Folding Stool
- Auto Detailing Masking Tape
- Wet & Dry Sandpaper 2 sheets each of 400, 600, 800 & 2000 grit (8 sheets in total)
- Trim Removal Tools
- Polythene Dust Sheet 3.5m x 2.6m (12ft x 9ft)
- Sharp Scissors or Blade
- Universal Paint Trigger (see Top Tip below)
- Face Mask
- Eye Protection
- Microfibre Applicator Pad

Top Tip Not all sandpaper is created equal. A good quality one will help you complete the project successfully, and a bad quality one, well, you don't want to go there - I *have*, and it's a depressing place! I prefer *Klingspor* and *Titan*; *both* will serve you well and at a reasonable price. Just make sure that whatever you buy is suitable for wet sanding.

Top Tip A Universal Paint Trigger makes paint spraying oh *so, so* much easier. The handle securely clips onto the top of any standard rattle can, and away you go! It hugely increases your control and precision when spraying and eliminates the dreaded *finger* ache associated with rattle can use. Mine came from China via eBay for £4 ($5.49) five years ago and is still going strong, hundreds of applications later. Just put 'Universal Paint Trigger' in Google, and they will pop up.

Detailing Liquids Needed:

- Washing-Up Liquid (Soapy Water Spray)
- All-Purpose Cleaner (APC) Spray (diluted 1:5)
- Isopropyl Alcohol (IPA) Spray (diluted 70:30)
- Tar & Glue Remover Spray
- *Gloss Clear Coat for Plastic – see note below
- Spray Wax
- Thick Carnauba Wax
- Microfibre Detergent

Top Tip Washing-Up Liquid is an excellent degreaser and is perfect for the initial cleaning of the headlights. Take one of your spray bottles, fill it with warm water, add a squirt of Washing-up Liquid and shake well – bingo, a dirt-cheap, turbocharged Soapy Water degreaser! This spray bottle has the bonus of double use for this project, which I will explain later.

*A word on Gloss Clear Coat Spray – this is the final sealant stage of your project and will give the restoration protection and durability.

You'll need a quality brand with *non-yellowing, UV-resistant* qualities; the brand I've used with success is *Rust-Oleum*.

Preparation

You know the drill as well as I do by now. Get those beers on ice ready for when you finish, and grab a bottle of water or Thermocafé mug of coffee to sustain you. Remove your rings and watch, and make sure you aren't wearing anything zippy or with buttons or poppers that could scratch your car while you're working up close to it. Position your car preferably with room to safely work around it. Put the Gloss Clear Coat spray can indoors to come up to room temperature; the spray paint does not work well if too cold or hot. Get your Headlight Restoration Kit assembled, load your Carry Caddy with your Spray potions, Microfibre Cloths, etc., and have your Caddy handy. Put your Kneeling Mats and Folding Stool in position if you're working down low.

Headlights in or out?

If your headlights are relatively easy to remove, put your Gloves on and go ahead and remove them from the car. It's a lot easier to work on them if they're off the car, ideally on a workbench, and you won't have to worry about accidentally damaging your paintwork when sanding around the lens edges.

I'm going to cover the process for working on them, both *on* and *off* the car, if you can't remove them. First up is the process when working with them *off* the car.

Note: A quick heads up to let you know the following pages contain some repetition of the necessary stages of the headlight OFF and ON the car processes. I'm aware some readers may prefer to skip directly to their preferred process.

7 Stages to Heroic Headlight *'OFF'* the Car Restoration

Stage 1

Detail the Headlight Aperture: Consult your handbook to see if you can remove the headlights. Another advantage of removing them is that you can detail both the bodywork light aperture and the back of the headlight.

Put the headlights somewhere safe; we'll come back to them in a bit. Grab your APC spray and generously spritz down the light aperture and allow the APC to soak for a minute.

Take a clean Microfibre Cloth and clean out all the grime, folding the Cloth as you go. Go in with your Exterior Detailing Brush if necessary. Give it a second hit of APC spray and wipe that down with the Cloth. If your Detailing Brush needs backup, you may find it helps to get into the tight spots by taking a flat-ended Trim Removal Tool, wrapping a Microfibre Cloth around the flat end, and using that to delve and clean in the tight areas. Finish by giving the aperture a spritz of spray wax and buff it with a clean Cloth. Repeat for the other headlight aperture, and when finished, move your APC, Wax, Trim Tool, Detailing Brush and used Cloths away from the car out of the way.

Stage 2

Give the Lights a Pre-Sanding Clean: Now to tackle those pesky headlights. Get them up somewhere where you can work in comfort to give them your best effort. You want to be sanding nice clean units, so a bit of attention here will serve you well:

Soapy Water Spruce: Take your Soapy Water Spray bottle and spray down the whole unit, front and back. Grab your Exterior Detailing Brush and work it over the back of the unit to clean out any grime. Then wipe everything down with a clean Microfibre Cloth and repeat on the other headlight.

APC Hit: Follow that by spraying the unit down with APC front and back, again working the Exterior Detailing Brush over the back area to banish any lingering grime. Wipe everything down with a clean Microfibre Cloth and repeat on the other headlight.

Tar & Glue Remover Attack: Call me fussy, but we're home detailers, right? So, finish off the pre-sanding clean-up by grabbing your Tar & Glue Remover Spray. Mist down the whole lens and allow it to soak for one minute, then wipe the lens down with a clean Microfibre Cloth and repeat for the other headlight.

Stage 3

Assess the Lens Condition: Now they're nice and clean, you can see what you're working with. If the headlights are *really* bad with significant cloudy hazing and severe pitting from stone chips, you'll need to start with *400 Grit* sandpaper. If the headlights have only moderate imperfections with a lesser degree of cloudy hazing and less serious stone chip pitting, you can start with *600 Grit* sandpaper. If you're unsure, run your un-gloved fingertips over the clean, dry lens surface and feel what your fingertips tell you. If you remain unsure, err on the side of caution, and start with 400 Grit, you won't hurt the lens because you're going to work your way through the grit ranges until you get a near-perfect finish!

Stage 4

Start Wet Sanding: Yes, we will *wet* sand these bad boys using the same Soapy Water spray you used to clean the units. That little squirt of washing-up liquid in the spray bottle is going to help give you great lubricity during the sanding stages.

A quick heads up before you start: you may find the first 400 grit pass a little nerve-wracking; the lens surface will begin to look much worse than before you started. But if you persevere and keep working through the grit ranges, as I explain, you'll arrive at a finish to be proud of.

400 Grit Sandpaper Step: It's up to you whether you work with gloves on or off during wet sanding; I tend to do this *without* gloves, as I like to be able to *feel* the surface changes as I go. Take the most abrasive of your sandpaper grades, the 400 Grit Sheet, and evenly fold it once and then again so the fold marks are in quarters. Cut the sheet into four equal quarters and set three pieces aside. Grab your Soapy Water Spray and spritz down the whole lens, then do the same on the piece of 400 Grit you're working with and make sure both lens and sandpaper are thoroughly wet. The Soapy Water acts as a lubricant and will prolong the life of the sandpaper.

Now, fold your piece of sandpaper again and start methodically sanding the lens using a circular motion. Cover every bit of the lens, applying moderate pressure – *don't* press down hard. Keep spritzing the lens with Soapy Water as you go, keeping it wet throughout the sanding. Make sure you sand right up to the *edges* and ends of the lens (the edges can be vulnerable to clear coat flaking if not carefully sanded). Keep going; this needs patience – you're going for a full, even, dulling down on the whole lens with the 400 Grit. You'll notice a milky texture liquid building up as you sand, this is normal, and it's just the damaged, oxidised surface of the lens coming off. When the sandpaper starts to lose its 'bite,' just re-fold it to expose a fresh side or grab a fresh piece and keep going.

When you have a full, even dulling down of the entire lens, including the edges and corners – *stop* with the 400 Grit and wipe the lens down with a clean Microfibre Cloth. You may notice that much of the *yellowish* oxidation has gone, and you're left with just the haze from your sanding efforts – lovely, repeat on the second headlight, then on to the next step. I warn you, the lens will look like a 'dog's dinner' right now, but don't worry, it's supposed to look like that at this stage.

600 Grit Sandpaper Step: Take your 600 Grit Sandpaper, fold and cut it in the same way and repeat the above process; the only change with the 600 Grit is you can afford to press down *slightly* harder. Don't forget to work the sandpaper right up to and along the *edges* and ends of the lens. I can't tell you how many light lens restoration and spray tinting efforts I've seen where the clear coat starts to flake at the edges due to inadequate prep! When the sandpaper starts to lose its 'bite,' just re-fold it to expose a fresh side or grab a fresh piece and keep going. As you come to the end of your 600 Grit pass, you'll see the lens has become noticeably smoother. Wipe the lens down with a clean Microfibre Cloth and brush your un-gloved fingertips lightly over the dry lens – see how much smoother it is? Repeat on the second headlight, then on to the next step.

800 Grit Sandpaper Step: You know the drill by now; fold and cut your 800 Grit in the same way. Spray down the lens and your 800 Grit with Soapy Water and get stuck in. I call this the first sanding *polish* stage! When the sandpaper starts to lose its 'bite,' just re-fold it to expose a fresh side and keep going. When you have finished going over the whole lens, including the edges and ends, wipe the lens down again. However, this time spray it down generously with APC before cleaning it off thoroughly with a clean Microfibre Cloth. If you don't do this, any sanding debris left on the lens will quickly clog the ultra-fine 2000 grit sandpaper on the next step, which will *reduce* the quality of the final sanding polish step. Once again, run your un-gloved fingertips lightly over the dry lens – it will *almost* but not quite be where you need it to be. Repeat on the second headlight, then on to the next step.

2000 Grit Sandpaper Step: Almost there, Driveway Warrior - this is the final sanding stage, and it's essentially a good sandpaper polish and will help give you a *crystal* clear, *gleaming* finish. Take your 2000 Grit Sandpaper, fold and cut it the same way. Spray down the lens and your 2000 Grit with Soapy Water and get polishing, folding the 2000 Grit as you go, once again using moderate pressure. Stick at it; this is your chance to make your lights super clear and ensure you cover every bit of the lens, including those all-important edges and ends. Remember to keep the lens thoroughly lubricated. When finished, you may notice tiny micro blemishes remaining here and there on the lens surface. Don't worry; it's impossible to make them 100% perfect, but the steps you've taken will take them as near perfect as possible. Later, you'll see the magic of the Gloss Clear Coat will fill and eradicate those tiny micro blemishes. Repeat on the second headlight, then on to the next step.

Stage 5

Final Clean Down: This is the final stage before you apply the sealant Gloss Clear Coat. Change to a *clean* pair of gloves.

Soapy Water Spruce: Take your Soapy Water Spray bottle and spray down the whole lens. Then wipe it dry with a clean Microfibre Cloth, and repeat on the second headlight.

Air Blast: Next, grab your Car Blower and give the whole unit front and back a thorough air blast to chase out any water hiding in the nooks and crannies. You don't want *any* moisture present while spraying the Gloss Clear Coat. Repeat on the second headlight.

IPA Bath: This will give your Gloss Clear Coat the best chance to bond fully to the lens. Grab your IPA Spray bottle and bathe the lens in IPA, getting it right on the edges and into the ends. The IPA removes any lingering grease or oils that may have been left behind from your hands, leaving the lens super clean for the clear coat spray. The IPA will flash away quickly with the help of your drying cloth. Take a *super* clean, fresh Microfibre Cloth, wipe the lens carefully down and repeat on the second headlight.

Stage 6

Gloss Clear Coat: Phew, we're on the final stage, don't mess up now; you're *soo* close to a *stunning* result! If possible, it's best to do the spraying in a shed or garage, with the door open for ventilation, so you're out of the elements.

Whether you're spraying them in the shed or outside, get the lenses propped securely up on a clean, stable surface, positioned *far* enough apart to avoid overspray settling on either of the units and for ease of spraying.

The Key to Successful Rattle Can Spraying

The key to successful paint spraying is the ambient conditions being favourable. Preferably warmish weather, but not humid.

If you're outside, it must be dry, with *no* breeze or dust. There must be *frenzied* and prolonged rattle can shaking – if it says to shake for two minutes, go mad and shake it for three minutes. Your technique will be crucial, and I *highly recommend* thoroughly reading the rattle can instructions and following them to the letter!

Shake It Like Dad at the Disco!

Ok, so now you can retrieve your Gloss Clear Coat rattle can from indoors and make sure it's at room temperature. Now start shaking it like you mean it. After a bit, you'll hear the mixing ball start to rattle – keep going for a full three minutes of shaking. When you have finished shaking, securely clip your Universal Paint Trigger onto the collar at the top of the rattle can. If you're working in *cold* conditions, make sure the Trigger Handle is at room temperature *before* pushing it onto the rattle can.

If the plastic around the Paint Trigger aperture is *too cold*, it will be brittle and may *crack* as you push it onto the rattle can – if that happens, the Paint Trigger will be useless.

Let's Get Spraying

The moment of truth, all your diligent, high-quality prep so far, brings you to this point, and you're going to *nail* this, my friend.

Before you start spraying, I want you to keep this phrase in mind and keep repeating it if you have to:

"Light, even coats."

"Light, even coats."

I cannot express just how *crucial* this is; once you get spraying, the *temptation* is to lay it on good and thick – but that will cause runs and *RUIN* your project. So, we're going to lay down three light, even coats, allowing each coat to dry before moving on to the next one.

First Coat: Check that the spray nozzle exit hole is right in the middle of the Paint Trigger aperture, so the edge of the aperture doesn't foul your nice smooth stream of paint spray when you squeeze the trigger. Do a quick test spray away from the lights to check that the spray comes out nice and evenly. If the spray stream looks good, hold the can 20-30cm (8-12in) away from the lens surface and lay down your first coat. Work from top to bottom and side to side, ensuring you get the top and bottom edges into the ends. You're looking for *no more* than 6-8 passes of the rattle can for this coat.

Don't worry if the surface looks a bit like 'orange peel' after the first coat, your second and third coats will sort that out, no problem! Allow the first coat to dry for 20 minutes or as per the rattle can instructions.

Top Tip To help prevent the nozzle clogging between coats, turn the rattle can *upside down* and spray well away from the lights until the nozzle ejects air only.

Second Coat: Give the rattle can another quick shake and check that the spray nozzle exit hole is still in the middle of the Paint Trigger aperture. Do a quick test spray away from the lights to check that the spray still comes out nice and evenly. Then lay down your second *light even coat* in the same way, doing no more than 6-8 passes. Allow to dry for 20 minutes or as per the rattle can instructions.

Third and Final Coat: Give the rattle can another shake for ten seconds and check that the spray nozzle exit hole is in the middle of the Paint Trigger aperture. Do a test spray away from the lights to check that the spray still comes out evenly. Then, lay down your third *light even coat*, in the same way, doing no more than 6-8 passes.

Stage 7

The Final Drying Stage: You are *DONE* – congratulations, Driveway Warrior; no doubt those lights look flipping magnificent! You don't have to do anything more at this stage – just get them inside somewhere safe and leave everything to dry for a *full* 24 hours.

Aside from that, please do not be tempted to *touch* them until the 24 hours is up. I know you want to get them back on the car, but don't do it, my friend; you'll ruin the awesome job you've put so much effort into. The drying stage will enable the layered up clear coat to cure and develop full optical clarity of finish – so you'll see how the true finish looks after 24 hours!

Clear Up

See Page 172 after the Headlight *ON* the Car Method.

24 Hours Later

So, by now, I bet they're looking almost new. But they don't just look *dazzling*; they're also semi-permanently protected against stone chip pitting, UV damage, and future yellowing or oxidation clouding – what an *EPIC* result!

A Final Wet Sand?

At this stage, we have two options:

1. The headlights already look awesome and if you're happy with the finish, leave them as they are.

2. If your detailing instincts have cursed you with *perfectionism*, you may notice an ever so slight 'orange peel' effect on the lens. Orange peel is the *bane* of spray paint, even for professional body shops, and can diminish the quality of the finish. So, if you want to go that *extra* step, you can give them one final polish with 2000 Grit to remove all traces of orange peel. Take a fresh piece of 2000 Grit, use plenty of soapy water spray to lubricate, and follow the same '2000 Grit Sandpaper Step' process as in Stage 4 on Page 158. Having completed that, you'll have satisfied all your perfectionist tendencies. Well done you!

Blissful Wax on

Whether or not you decide to do a final wet sand, the last stage is to apply a layer of old-school thick Carnauba Wax to seal in the finish. Grab your favourite *Wax and a Microfibre Applicator Pad and lay down a lovely satisfying layer of Wax. Let it haze up for a minute, then buff with a clean Microfibre Cloth to a super sexy shine - pure bliss! You can now carefully put the freshly detailed and restored headlights back on the car.

*I love the cheap and cheerful *Simoniz Original Carnauba Wax,* which comes in a cool 150g gold and red tin. It may be inexpensive, but it's a doddle to use and delivers flipping *cracking* results for around £9 ($12).

Part 2

9 Stages to Handsome Headlight *'ON'* the Car Restoration

Ok, so some headlights are a complete pain in the *wotsit* to remove. When I wanted to DIY cure one of my VW Golf headlights from filling with water every time it rained, a local mechanic took *three* flipping hours to remove the headlight for me. Maybe he thought I was paying him by the hour, hmm!

Anyway, we have a procedure to sort them out while still on the car, and they'll look just as *awesome* as if you'd taken them off the car. If it's possible to move the car into a garage, out of the elements, do it now as it's easier to use rattle can spray paint when you don't have rain, dust, wind, or any other weather elements to contend with. If you're working outside, it's a case of checking the weather forecast in advance and keeping your fingers crossed for a warm, dry day with no breeze!

Stage 1

Give Them a Pre-Sanding Clean

You want to be sanding nice clean units, so a bit of effort here will serve you well and ultimately help you toward a stunning result.

Soapy Water Spruce: Take your Soapy Water spray and spritz down the whole unit, including the edges where it meets the bodywork and the bodywork area immediately around the headlight. This will help the Stage 2 masking tape stick to the paintwork where you want it to stick. Then, take a flat-tipped Plastic Trim Tool, wrap a Microfibre Cloth around the flat tip, and use that to gently clean in between the headlight and the bodywork. Next, wipe everything down, including the surrounding paintwork, with a clean, dry Microfibre Cloth. Repeat on the other headlight.

> *Top Tip* Depending on the design of your headlights, you may get improved access to them by raising the bonnet (hood).

APC Hit: Follow that by spraying the unit down with APC. Once again, work the Trim Tool wrapped with the Cloth to gently clean in between the headlight and the bodywork. Wipe everything down with a clean Microfibre Cloth and repeat for the other headlight.

Tar & Glue Remover Attack: Call me fussy, but we're home detailers. So, finish off the pre-sanding clean-up by grabbing your Tar & Glue Remover Spray. Mist down the whole lens and allow it to soak for one minute, then wipe the lens down with a clean Microfibre Cloth. Make sure everything, including the immediate surrounding paintwork, is dry. Repeat for the other headlight.

Stage 2

Masking - Protect the Paint from Sandpaper: Next, we need to mask right around the edge of the whole headlight on the paintwork to protect the paint during the sanding stages. So, take your 3M Auto Detailing Masking Tape, using strips approximately 15cm (6in) long, and start to tape all around the lens using overlapping pieces of Tape.

Make sure there are no exposed areas of paintwork between the lens edge and masking tape. I like to keep going until I have a good 10cm (4in) barrier of masking tape barrier around the whole lens to keep the paintwork *super* safe from any over-enthusiastic sanding! Finish by going around all the tape, pressing it well and truly *down* with your thumb.

Stage 3

Assess the Lens Condition: Now they're nice and clean, you can see what you're working with. If the headlights are *really* bad with significant cloudy hazing and severe pitting from stone chips, you'll need to start with *400 Grit* sandpaper. If the headlights have only moderate imperfections with a lesser degree of cloudy hazing and less serious stone chip pitting, you can start with *600 Grit* sandpaper. If you're unsure, run your un-gloved fingertips over the clean, dry lens surface and feel what your fingertips tell you. If you remain unsure, err on the side of caution, and start with 400 Grit. You won't hurt the lens because you're going to work your way through the grit ranges until you get a near-perfect finish!

Stage 4

Start Wet Sanding: Yes, we will *wet* sand these bad boys using the same Soapy Water Spray you used to clean the units. That little squirt of washing-up liquid in the spray bottle is going to help give you great lubricity during the sanding stages.

400 Grit Sandpaper Step: It's up to you whether you work with gloves on or off during wet sanding; I tend to do this *without* gloves, as I like to be able to *feel* the surface changes as I go. Follow the same steps shown in Stage 4 on Page 158 for the '400 Grit Sandpaper Step.'

600 Grit Sandpaper Step: Take your 600 Grit Sandpaper, fold and cut it the same way and repeat the same steps shown in Stage 4 on Page 159 for the '600 Grit Sandpaper Step.' The only change with the 600 Grit is that you can afford to press down *slightly* harder.

800 Grit Sandpaper Step: You know the drill by now. Fold and cut your 800 Grit in the same way and repeat the same steps shown in Stage 4 on Page 159 for the '800 Grit Sandpaper Step.'

2000 Grit Sandpaper Step: You're almost there, Driveway Warrior. This is the final sanding stage, and it's essentially a good sandpaper polish and will help give you a *crystal* clear, *gleaming* finish. Repeat the same steps in Stage 4 on Page 159 for the '2000 Grit Sandpaper Step.'

Stage 5

Second Quick Clean Down: This is the penultimate cleaning stage before you apply the sealant Gloss Clear Coat. Change to a *clean* pair of gloves.

Soapy Water Spruce: Take your Soapy Water Spray bottle and spray down the whole lens. Then wipe it dry with a clean Microfibre Cloth and repeat on the second headlight.

Air Blast: Next, grab your Car Blower and give the whole unit a thorough air blast to tease out any water hiding in the nooks and crannies between the lens and the bodywork. You don't want *any* moisture present while spraying the Gloss Clear Coat. Repeat on the second headlight.

IPA Bath: This will give your Gloss Clear Coat the best chance to bond fully to the lens. Grab your IPA Spray and bathe the lens in IPA, getting it right on the edges and into the corners. The IPA removes any lingering grease or oils that may have been left behind from your hands, leaving the lens super clean for the Clear Coat Spray. The IPA will flash away quickly with the help of your drying cloth. Take a *super* clean fresh Microfibre Cloth, wipe the lens carefully down and repeat on the second headlight.

Stage 6

Protect the Paintwork from Overspray: If you've ever done any rattle can spraying outdoors before, you'll know that the fine mist coming out of the nozzle can go *everywhere* but where you want it to go, even with just the *gentlest* of summer breeze wafting by. This means that overspray getting on your paintwork is a massive risk, particularly if you're working outside in the elements. So, if you want to avoid a *mammoth* Clay Mitt or Clay Bar session to remove overspray from your paintwork, really go to town on thoroughly covering everything up – here's how:

Lay own the Plastic Sheet

With your Gloves off, take the large piece of Polythene Dust Sheet, and spread it right out over the whole front of the car, going up to the front edge of the roof. Let it hang down loosely over the windscreen and bonnet (hood), down over the headlights and front bumper to the ground. Now you can see another reason why it's impossible to do this if it's windy!

Cut-Stick-Press

Once you have the sheeting roughly laid out in position, hanging down loosely from the top near the front of the roof, take your 3M Auto Detailing Masking Tape and tape the top of the sheet to the top of the car, near the front of the roofline. Use several pieces of tape along the width of the roofline. Then, *gently* pull the plastic so it's a bit tighter and tape all around the sides so it's secured to the car at the top and both sides.

Cut It: Now, take your sharp scissors and carefully cut around the edge of the headlight lens. You want to try and cut the lens access hole a little too *small*, as this will enable you to tape down the excess plastic securely down to the edges of the headlights. Repeat for the other side, and when you have finished cutting, move the scissors out of the way.

Stick It: Once you have the hole cut out roughly to the shape of the lens, take your Masking Tape and tape the plastic down right around the headlight, applying the same method used earlier in Stage 2. *Remember* that you have already carefully taped around the lens in Stage 2, so all you're doing here is securing the plastic well and truly.

Press It Down: Once again, go around the tape, pressing it down with your thumb, making sure it's good and stuck.

Stage 7

Final IPA Spritz: We need to do *one* more IPA spray down to make sure there's nothing on the headlights from your hands or elsewhere from the plastic sheeting *antics*. Change to a clean pair of gloves, grab your IPA Spray bottle, shake it and spritz down the lens. Get it right on the edges and into the ends, and wipe it dry with a super clean, dry Microfibre Cloth. Repeat on the other headlight and move your IPA spray bottle away from the car out of the way.

Stage 8

Gloss Clear Coat: Phew, we're on the final stage, don't mess up now; you're *soo* close to a *stunning* result!

The Key to Successful Rattle Can Spraying

The key to successful paint spraying is favourable ambient conditions - preferably warmish weather, but not humid. If you're outside, it must be dry, with *no* breeze or dust, and there must be *frenzied* and prolonged rattle can shaking. If it says to shake for two minutes, go mad and shake it for three minutes. Your technique will be crucial, and I *highly recommend* thoroughly reading the rattle can instructions and following them to the letter!

Shake It Like Dad at the Disco!

Ok, so now you can retrieve your Gloss Clear Coat rattle can from indoors and make sure it's at room temperature. Now start shaking it like you mean it. After a bit, you'll hear the mixing ball start to rattle - keep going for a full three minutes of shaking. When you have finished shaking, securely clip your Paint Trigger onto the collar at the top of the rattle can.

If you're working in *cold* conditions, make sure the Paint Trigger is 'room temperature' *before* pushing it onto the rattle can. If the plastic around the Paint Trigger aperture is *too cold*, it will be brittle and may *crack* as you push it onto the rattle can – if that happens, the Paint Trigger will be useless.

Let's Get Spraying

The moment of truth, all your diligent, high-quality prep so far, brings you to this point, and you're going to *nail* this, my friend.

Before you start spraying, I want you to keep this phrase in mind and keep repeating it if you have to:

"Light, even coats."

"Light, even coats."

I cannot express just how *crucial* this is; once you get spraying, the *temptation* is to lay it on good and thick – but that will cause runs and *RUIN* your project. So, we're going to lay down three light, even coats, allowing each coat to dry before moving on to the next one.

First Coat: Check that the spray nozzle exit hole is in the middle of the Paint Trigger aperture. You don't want the aperture edge to foul your paint spray stream when you squeeze the trigger. Do a quick test spray away from the lights to check that the spray comes out evenly. If the spray stream looks good, hold the can 20-30cm (8-12in) from the lens surface and lay down your first coat. Work from top to bottom and side to side, ensuring you get the top and bottom edges into the ends. You're looking for *no more* than 6-8 passes of the rattle can for this coat.

Don't worry if the surface looks a bit like 'orange peel' after the first coat, your second and third coats will sort that out, no problem! Allow the first coat to dry for 20 minutes or as per the rattle can instructions.

Second Coat: Give the rattle can another shake and follow the same steps as the 'First Coat' above.

Third and Final Coat: Give the rattle can another shake for ten seconds and follow the same steps as the 'First Coat' above.

Stage 9

The Final Drying Stage: You are *DONE* – congratulations, Driveway Warrior; no doubt those lights look flipping magnificent! PLEASE do not be tempted to touch them until 24 hours is up. You don't have to do anything more at this stage; put the car under cover if possible and leave everything to dry for a *full* 24 hours. The drying stage will enable the layered up clear coat to cure and develop full optical clarity of finish – so you'll see how the true finish looks after 24 hours!

24 Hours Later

So, by now, I bet they're looking almost new. But they don't just *look dazzling*; they're also semi-permanently protected against stone chip pitting, UV damage, and future yellowing or oxidation clouding – what an *EPIC* result!

A Final Wet Sand?

At this stage, we have two options:

1. The headlights already look awesome and if you're happy with the finish, leave them as they are.

2. If your detailing instincts have cursed you with *perfectionism*, you may notice an ever so slight 'orange peel' effect on the lens. Orange peel is the *bane* of spray paint, even for professional body shops, and can diminish the quality of the finish. So, if you want to go that *extra* step, you can give them one final polish with 2000 Grit to remove all traces of orange peel. Take a fresh piece of 2000 Grit, use plenty of soapy water spray to lubricate, and follow the same '2000 Grit Sandpaper Step' process as previously done in Stage 4 on Page 159. Having done so, you'll have satisfied all your perfectionist tendencies. Well done you!

Blissful Wax on

Whether or not you decide to do a final 2000 grit wet sand, the final stage is to apply a layer of old-school thick Carnauba Wax to seal in the finish. Grab your favourite *Wax and a Microfibre Applicator Pad, and lay down a lovely satisfying layer of wax. Let it haze up for a minute, then buff with a clean Microfibre Cloth to a super sexy shine – pure bliss! Carefully remove the tape and plastic sheet from the car and discard them.

*I love the cheap and cheerful *Simoniz Original Carnauba Wax*; see Page 163.

Clear Up

On to a quick clear up to keep all your kit in first-rate order. Start by returning all your Liquids, Wax, Rattle can and Spray bottles to storage, but do first check all lids and spray heads are secure and keep an eye on any dilutables that need topping up.

Grab your Kneeling Mat, Stool, Car Blower, Extension Lead, Trim Tools and Paint Trigger and give them a quick wipe down if necessary and store them away. Keep your leftover sandpaper and detailing tape somewhere safe. Give your exterior detailing brush a squirt with washing-up liquid, work the suds in with your gloved fingers, and give it a good rinse under a hot tap, then give it a wrist flick and put it somewhere to dry.

Gather your used Cloths and Applicator Pad and pop them in the washing machine at 40°C (104°F), add the Non-Bio Delicates Detergent and stick on a 30-minute cycle with a gentle spin. When they come out of the machine, give them a good shake to re-set the fibres and hang everything on a clothes dryer to dry naturally, using a peg for the Applicator Pad. The next day when your Cloths are fully dry, store them away safely.

Reflection

No more *squinty-eyed* night driving for you, my friend; those *crystal clear* lens babies will now illuminate the way like a *lightning bolt* carving through the darkness! Particularly if you upgrade the stock bulbs to something more high performance. For the cost of a few sheets of sandpaper, plus your elbow grease, you've saved *hundreds*, if not *thousands*, on new headlights. In the case of my Cayman, I saved just over £1000 ($1342) on the cost of new units. Oh, and a pro detailer would have charged you around £144 ($194) for the work you've done so awesomely.

With clarity of *vision* comes clarity of *thought*. In this Driveway Warrior's case, the clarity of thought is the realisation of the *epic* foolishness of my attempting to negotiate the pitch-black Surrey wilderness with shamefully hazed up, cloudy headlights. *Doh,* my bad!

On to the next detailing project, friend - what the devil could we tackle next? Aha, I know! Follow me for the delights of *Intimate Undercarriage Detailing & Protection* - read on in the next chapter, Driveway Warrior ...

CHAPTER 10
Ultimate Undercarriage Detail & Protection

Part 1 - Epic Under Trays Detailing

Thinking about an intro for this un-sexy, though thoroughly relevant detailing project, I was forced to go with the lowest common denominator – pants referencing! Here goes …

Much like a pair of *under-crackers* (men's underpants for non-UK readers), it's a good idea to keep your car's undercarriage freshly detailed and free from soiling! Fortunately, unlike your under-crackers, you won't need to detail your undercarriage daily. You certainly won't have to endure social judgment and alienation if you don't detail your undercarriage regularly!

Why Bother? – No one will See It

"But, why on earth bother with detailing the underbody?" I hear you cry. "No one will ever *flipping* well see it! We've spruced up the engine bay so that you could eat your dinner off it. We've made those wheel wells positively *gleam*, so isn't detailing the undercarriage taking things just a step too far?"

Absolutely-*blinking*-well not, my friend! Remember, it's all in the *detailing*. The fact is, if left unchecked, the build-up of dirt, grime, salt and debris on the underbody will start to corrode your car from the bottom up.

That's the onset of structural or chassis issues right there, and there are components of the underbody that are key to vehicle safety. Allowing dirt and grime to accumulate over a prolonged period can negatively impact safety while eroding the car's value and reducing its useful life.

Is It a One-Off Project?

If, like me, you no longer drive your prestige motor during the winter months, you'll probably only need to do this project once. I'll be honest; my daily driver VW Golf doesn't receive the same detailing love that my Porsche Cayman does. It's the Golf that's punished on our autumn and winter roads, while the Porsche stays cosy under a Halfords outdoor breathable car cover from October to March. Assuming you don't regularly off-road your ride, you absolutely can bling up and protect your underbody.

Not a Cloud in the Sky

Most of the advice given in previous chapters about detailing in ideal weather conditions applies to your Ultimate Undercarriage & Protection Project, with a few exceptions. May I refer you to Chapter 1, *'Sunshine – the Safe Wash Enemy.'* You shouldn't normally start detailing in direct sunlight if you want to avoid issues with your detailing liquids drying out too quickly on your car and screwing up the quality of your finish. However, you're detailing under the car in this case, so it would help to pick a couple of days when the forecast is dry and preferably warm. A bit of spring or summer warmth in the air will help the underbody dry out after you've cleaned it.

Undercarriage Detailing & Protection Kit List

You're going to *need* a Pressure Washer for this one; a Hosepipe and Nozzle alone doesn't have the power to do it properly.

Detailing Hardware Required:

- Carry Caddy
- Nitrile Gloves
- Kneeling Mats
- Folding Stool
- Pressure Washer Kit inc. Snow Foam Dispenser & Extension Cable
- Wheel Wash Bucket & Kit Brushes
- *Extra Long Handle Wash Brush
- Car Blower
- Detail Guardz

- Microfibre Cloths (a bag full of downgraded cloths)
- Universal Paint Trigger (see Page 154)
- Eye Protection
- Face Mask
- Rhino Ramps (see Page 184)
- Wheel Chocks x 2
- **Tarpaulin 2 x 3m (6.5 x 9.8ft)
- Spare Bucket
- Plastic Toothbrush
- Small Container with Lid
- Food bag
- Auto Detailing Tape
- LED Light
- Portable Small Flatbed Fan Heater

*The extra-long handle brush I used for the underbody was the Kent Car Care Dip & Wash Brush. A mighty 55.4cm (22in) long and only 500g (17.6oz) in weight, it's perfect for reaching under to scrub muck away from the undercarriage (but don't use it on your paintwork). It is available online for approximately £7.50 ($10.17).

**Tarpaulin size available from Screwfix for £5.99 ($8.15).

Detailing Liquids Needed:

- Rust Repellent (diluted 1:3)
- Pure Shampoo
- Snow Foam
- All-Purpose Cleaner (APC) Spray (diluted 1:4)
- Isopropyl Alcohol (IPA) Spray (diluted 70:30)
- Perl Dressing Spray (diluted 1:3)
- WD-40
- Tar & Glue Remover Spray
- *Hammerite Waxoyl Original Clear Spray 400 ml x 5 cans
- Microfibre Detergent

*Hammerite Waxoyl Original Clear is an unmistakable yellow can containing a thick, waxy fluid saturated with a powerful rust killer. When sprayed on metal, it *chases out moisture and forms a flexible, weatherproof skin that doesn't crack, dry out or wash off in the rain.*

It is available online for approximately £11 ($14.90) for 400ml.

Splitting Up the Work

I find it easier to split up the work in this project over two days. So, on day 1, I detail and protect the five under trays plus the fixings. Then, on day 2, I concentrate on doing a cracking job on the undercarriage.

Day 1

Getting Started - 9 am

The Cayman has five underbody plastic trays protecting the components and supporting airflow under the car at speed. I didn't fancy in the slightest getting the car up on axle stands and lumbering around on my back, struggling to remove a couple of dozen Torx bolts. So, bright and early, I chucked the tarp (from the above hardware list) in the car and took a rapid spin over to my obliging local Porsche indie, Golding Barn Garage. While they whipped off the panels sharpish so I could take them home to detail, along with the underbody, I spread the tarp out over the Cayman rear cargo area to protect everything.

When the trays were removed, I was surprised by the *vast* amount of debris and grit that fell out of them. It struck me that this would have been sitting in the trays holding moisture, which would start to cause corrosion problems on the underbody. Having seen all that damp crud falling out and covering the floor of the lifting bay, I was *super* glad I'd opted to do this 'under trays off' detailing and protection project.

By some miracle, all five mucky plastic trays, plus a plastic bag full of fixings, fit snugly in the rear cargo area – a testament to the *awesome* versatility of the Cayman, methinks! The cost to remove the panels and replace them after detailing was £45 ($61) – pretty reasonable considering what a pain they would be to deal with on the driveway.

Your car may or may not have similar underbody panels to deal with, so before you get started, take a gander underneath to see what's what before you plan your project.

Given the now unprotected underbody, the short trip home from the garage was rather more *sedate* to ensure nothing delicate would be damaged through hooligan driving antics!

6 Steps to Terrific Under Trays Bling

I found the easiest way to tackle the five large under trays was to lay them all down flat in the garden. Get your Under Tray Detailing Kit assembled and load your Carry Caddy with your Spray potions, Microfibre Cloths, Wheel Detailing Brushes, Pressure Washer, Snow Foam, etc., and have your Caddy handy. Take your Wheel Bucket with Grit Guard in the bottom, add two capfuls of Pure Shampoo, fill with hot water and swirl the water around to mix in the shampoo.

Next, take the bag of fixings – there should be a mixture of metal and plastic screws. Take the Small Container, dump all the fixings in, and fill it with hot soapy water from the Wheel Bucket. Put the lid on, give it a quick shake, and set it aside to soak. We'll come back to it later. Don't forget to grab a bottle of water or a Thermocafé mug of coffee to sustain you while you work.

Step 1: Pre-Soak with Snow Foam: With your Gloves on, load up your Pressure Washer Cannon with Snow Foam 1:4 mix (snow foam to tap water), and generously coat one side of all five trays with snow foam. Allow the Snow Foam to soak for two minutes, then give each tray a thorough Pressure Washer rinse with water. Turn the five trays over and repeat on the other side.

Step 2: Brush Attack with Snow Foam: For the second time, generously coat one side of all five trays with Snow Foam and allow it to soak for two minutes. If possible, get one of the trays up on a workbench at this point so that you can brushwork it more comfortably. If there is no workbench, use your Kneeling Mats and Stool if you wish. Get scrubbing with the ValetPro Chemical Resistant Wheel Brush. Work the brush thoroughly over one side of the whole tray, rinsing and lubricating the brush regularly in your wheel wash bucket as you go. Repeat on the other four trays.

Then, turn all five trays over and repeat the Snow Foam and Brush treatment on them. Finish this stage by giving each tray a thorough pressure washer rinse down with water on both sides.

Step 3: Tar & Glue Remover Spray: This is the *penultimate* cleaning stage and will tackle the remaining tar spots that will have been flung up from the road and peppered the bottom of the trays. Eye Protection on, take your Tar & Glue Remover Spray, shake well and generously spray down one side of all five trays and allow to soak for two minutes. Take your ValetPro Chemical Resistant Wheel Brush, and once again, work the Brush over the tray. Rinse and lubricate the brush regularly in your Wheel Wash Bucket as you go, and then repeat on the other four trays. Next, turn all five trays over and repeat the Tar & Glue spray and brush treatment on them all. Finish this stage by giving each tray a thorough pressure washer water rinse on both sides. When finished, move your Tar & Glue Remover, Wheel Wash Bucket, Brushes, Pressure Washer, and Snow Foam Lance away from your working area out of the way.

Step 4: IPA Wipe Down: We're at the final cleaning stage - phew! Lean the trays up somewhere if you can, grab your Car Blower, and spend a few minutes giving the trays a good air blast on both sides to expel the rinse water out of the nooks and crannies.

Then, to prepare the trays to receive the dressing protectant, put the trays down flat again. Take your IPA Spray diluted at 70:30, shake well and generously spritz all five trays on one side with IPA, getting the spray into all the tight spots. Next, grab your bag of downgraded Microfibre Cloths, wipe down the whole area with a clean Cloth, drying it as much as you can, folding the Cloth as you go.

You should find that your four detailing stages thus far have successfully cleaned the trays beautifully, and your Microfibre Cloth should be coming up satisfyingly clean. Most likely, it'll be the *cleanest* the trays have been since the car left the factory! Turn all five trays over and repeat the IPA Spray and Cloth wipe down. When you have finished, move your IPA Spray and used Cloths away from your working area out of the way.

Step 5: Dressing Solution Protection: Now for the thoroughly *satisfying* bit where the fruits of your labour start to reveal themselves. Take your CarPro Perl Dressing Spray, diluted 1:3, shake well and spritz the Perl all over one side of the trays. Then, take a fresh, downgraded Microfibre Cloth and work the Perl Dressing in well. Fold the Cloth as you work the dressing into the nooks and crannies, not forgetting the edges. Turn the trays over and repeat on the other side. As well as the *magnificent* restorative, mid-sheen darkening of the finish, the Perl also gives the plastics welcome UV and weathering protection.

Due to the amount of abuse the under trays routinely receive from day-to-day driving, I recommend doing *two* coats of Perl Dressing. Allow the first coat of dressing to cure for ten minutes, then lay down a second coat of Perl and work it in again with the Microfibre Cloth and repeat on the other side. You'll notice an even deeper, richer darkening and sheen of the plastics after the second coat – lovely! Move your Perl Dressing Spray and Cloths away from your working area out of the way. *Under Trays done* – detailed and protected! Stand back and admire your results now, as you won't get much of a chance to see them when they're back on the car! Lean them up on their sides, preferably under cover, or throw an old tarp over to keep them clean before they go back on the car.

Step 6: Gleam Clean the Fixings: *Oh*, you are *soo* a home detailer now, my friend – anyone who details fixings and screws has to be by definition!

Retrieve the Container of fixings, WD-40, IPA Spray, Perl Dressing Spray and Toothbrush and dump the dirty water out of the fixings Container. Get the fixings up somewhere where you can work on them in comfort. Grab the IPA Spray, generously spritz all the fixings down with IPA and go to work on them with the Toothbrush.

Next, dry them off with a downgraded Microfibre Cloth and separate the metal from the plastic fixings. Give the metal screws a quick hit with WD-40 to give a bit of corrosion protection, and dump them in a clean food bag. Finally, take the Perl Dressing Spray, give the plastic fixings a quick spritz, and wipe them down again with a clean Microfibre Cloth. Then dump them in the bag with the metal screws. *Fixings are detailed* and worthy of attaching to your shiny under trays! Put the bag in the car, ready to go.

Day 1 end

Mini Clear Up

We need to do a mini clear up to prep for day 2. Discard all the bucket water and give your Wheel Bucket and Grit Guard a rinsing blast with the Pressure Washer but hold back the Wheel Bucket as you'll need it in a bit. Next, gather all your used Cloths and chuck them in the washing machine, but *don't* wash them until the end of day 2. Store all your Spray bottles in the Caddy so they're ready to go.

Dump your Brushes in the Wheel Bucket and give them a quick rinsing blast with the Pressure Washer. Discard that rinse water in the water butt if you have one. Add a squirt of washing-up liquid to the bucket, half-fill it with hot tap water, swirling it around to mix. Spend a couple of minutes with your gloved hands in the soapy water, massaging the brushes clean. Then dump the dirty soapy water, give the brushes a final rinsing blast in the bucket with the pressure washer and discard that rinse water in the water butt.

Give the Brushes a good wrist flick to fling off the excess rinse water and lay them on an old towel under the radiator to dry or in the sun if you're lucky. Give your Wheel Bucket a quick rinse if needed and wipe it dry. Give your Pressure Washer and accessories, including the power cable, a quick wipe down to dry, and you're good to go for day 2!

Part 2 - Uber Undercarriage Detailing

Day 2 Preparation

Driveway Warriors all know the routine by now. The only change I'd like to see is to substitute those beers in the fridge for a nice bottle of Champers or another fizz of your preference. Plus, why not order your favourite pizza when you finish, in celebration of arriving at the *final* detailing project in this book? Go on, *treat* yourself, my friend; you more than deserve it! Grab a bottle of water or a Thermocafé mug of coffee to sustain you. Remove your rings and watch, and make sure you aren't wearing anything zippy or with buttons or poppers that could scratch your car while you're working up close to it.

Get your Undercarriage Detailing Kit assembled, retrieve your Pressure Washer and accessories, Snow Foam, Wheel Bucket, Extra-Long Handle Brush, Caddy with your Spray potions, etc., and have your Caddy handy. Position your car on a level, flat, stable surface (please no sloping driveways for this project), preferably with room to safely work around it.

> ***Top Tip*** An extremely useful and pleasant resource I deployed for this project was my 12-year-old niece, Lilly. She had a splendid time getting down on the ground by the side of the car to blast underneath with the pressure washer and squirting *Waxoyl* everywhere while I issued crucial instructions from my stool! It was the first time she'd ever *earned* pocket money, so it was a special moment, and she was delighted with her fee! Good company too - an assistant is optional, of course!

Prep the Waxoyl

Fill the spare Bucket with hot (not boiling) tap water, check the lids are firmly on the five Waxoyl cans, pop them all in the hot water and leave them there until you need them. I've found that the Waxoyl Spray works *far* better when it's warmed up, so unless you live in a warm region, the hot water bucket is the easiest way to achieve this! When you're ready to use the spray can, the water will have cooled to warm, and the Waxoyl will be *good to go*.

Car Ramps

Rather than raising the car on axle stands for this project, I find it much *quicker* and *easier* to get one end of the car up on ramps to aid access to the underbody. Detail the raised half of the underbody, then get the other end of the car on the ramps and repeat the detailing process on the other raised half. I found my detailing work wasn't overly hampered by not removing the wheels or wheel well liners; not entirely ideal, granted, but the ramps were an acceptable compromise.

What About Sorting the Wheel Arches?

Good point! The wheel arches form part of the underbody, and I cover the *Wicked Wheels off Wheel-Arch Detailing and Protection* in Chapter 5.

A Word on Rhino Ramps

I like Rhino Ramps because they're extra wide. They have a rugged, non-skid tread pattern base with a 17-degree incline – crucially giving easy access for low clearance sports cars. They're surprisingly light considering the 3,628 kg (8,000 Lbs) vehicle weight capacity per pair. Importantly, at W30 x L90 x H17cm (W12 x L35.5 x H6.7in), they are comfortably broad enough to accommodate the wide Cayman wheels. As a bonus, they stack neatly inside each other for convenient storage. I look at this kit as an investment, as I use them time and again for detailing and maintenance on the Cayman and Golf daily driver. They are available for approximately £75 ($101).

Raise the Front Up

Position the ramps in front of the front tyres, then get down low and line them up carefully. Ensure the tyres are centred in the middle of the ramps, with *no* overhang to the sides. This is where the generous width of the Rhino

Ramps helps you big time. When you're happy you have them lined up perfectly, give the ramps a good kick to help wedge them under the tyres before you drive carefully up the ramps.

Next, drive the car *slowly* and *steadily* up the ramps until you feel the car bump softly against the top resting point at the back of the ramps. This experience can be a little nerve-wracking the first time you do it, though rest assured, these ramps are superbly designed and safe if used correctly. After doing this a few times for your detailing projects, you'll be going up there with your eyes shut!

Remove the ignition key, put the handbrake firmly on and place the car in first gear. If you place the car in park (for an automatic gearbox/transmission), it will only lock the two rear wheels. Hop out of the car and check that you're happy with the position of the wheels on the ramps. Place your Detail Guardz on the outer edges of the rear tyres, then, as an extra safety precaution against the car rolling, carefully chock the two rear wheels that remain on the ground.

Check the Car's Stability

Call me fussy, but we need to do one final safety check before cracking on with some detailing. Now, gird your loins, and standing off to the side at the raised end, give the car a firm shove to check if it's super stable on the ramps. If the car doesn't budge, you're good to go. But if the car moves at all, you do *not* have it correctly raised. It's far better that the car falls off the ramps while you're shoving it from the side than it falls on any part of you while you're working on it!

Protect the Driveway

You don't want the Waxoyl to mess up your driveway, so we're going to protect the drive. If you're lucky enough to have a junior assistant as I did, get them to help you fold the tarp and spread it out under the car.

*It's important to note that we *don't* want the tarp *under* the ramps, as this would create a potential ramp slip hazard. The ramps need to be directly on the drive's surface, not on the tarp. When you have the tarp spread out, weigh it down at the corners with bricks or something similar.

Look After Your Knees

Grab your six Kneeling Mats, and place three interlocked mats on either side of the car for a *degree* of comfort when working down low. Either your knees or your young assistant will thank you!

6 Stages to Ultimate Undercarriage Detailing

Then, take your Wheel Bucket with Grit Guard at the bottom and add two capfuls of Pure Shampoo plus two capfuls of APC for extra foaming and cleaning power. Fill with hot water and swirl the water around to mix. Take your Auto Detailing Tape and run a tape strip on both sides of the car from the mid-point at the bottom sill, going up the side of the bodywork. This marker will help you see where to work under the raised half of the car when you're down low on the ground.

Stage 1: Rust Repellent & First Rinse: Don your Gloves, and as you'll be working down low at eye level, please put your eye protection on, too, as you don't want any of these chemicals splashing in your eyes. Grab the Rust Repellent Spray, shake well, and give all four brake discs (rotors) a good spray down. Allow it to cure for two minutes, and move the Rust Repellent away from the car out of the way.

Next, get down and give the undercarriage a thorough Pressure Washer water rinse down. Ensure your Pressure Washer attachment is the wide fan nozzle and keep the powerful spray moving. This will loosen and dislodge the worst of the muck under there. You'll need to move between the two sides of the car to cover the working area during all these cleaning stages.

Stage 2: Pre-Soak with Snow Foam: Grab your Snow Foam Dispenser and load up a good clingy 1:2 mix (Snow Foam to tap water). Shake well, then generously coat the undercarriage working area with a lovely *thick* Snow Foam blanket. Allow it to soak for five minutes, then rinse it down thoroughly again with the Pressure Washer.

Stage 3: Brush Attack with Snow Foam: Now, for the second time, generously coat the undercarriage working area with Snow Foam and allow it to sit for two minutes. Then, take your Wheel Wash Bucket and Extra-Long Handle Brush, get down low on your mats, reach under and get scrubbing everywhere under there as best you can, rinsing the Brush regularly in your Wheel Wash Bucket. When finished with the Brush on this stage, give the whole working area another thorough rinse down with Pressure Washer water. Move your Snow Foam cannon away from the car out of the way.

Stage 4: Tar & Glue Remover – 2 Hits: This is the *final* cleaning stage in your working area and will help tackle tar spots that will have flung up from the road and peppered the whole undercarriage. Take your Tar & Glue Remover Spray, shake well and *generously* spray down the whole undercarriage working area and allow it to soak for two minutes. Then, take your Wheel Wash Bucket and Extra Long Handle Brush once again, and get scrubbing everywhere under there, rinsing the brush regularly in your Wheel Wash Bucket. When finished with the Brush on this stage, give the whole working area another thorough rinse down with the Pressure Washer.

Next, for the second time, take your Tar & Glue Remover, *generously* spray down the whole undercarriage working area and allow it to soak for another two minutes. For the final time on this working area, take your Wheel Wash Bucket and Extra Long Handle Brush, and get scrubbing everywhere under there, rinsing the brush regularly in your Wheel Wash Bucket. When finished with the Brush on this stage, give the whole working area a final thorough rinse down with Pressure Washer water. Move your Tar & Glue Remover, Wheel Wash Bucket, Brush and Pressure Washer away from the car out of the way.

Let's Get the Back Half Done

We need to raise the back of the car to detail the back half of the undercarriage. So, remove your Detail Guardz and Wheel Chocks from the rear wheels and pull the tarp away from the car. Then, slowly and carefully reverse down the ramps and turn the car around. Next, follow the same process as before, *slowly* and *steadily* reverse up the ramps until you feel the car bump *softly* against the resting point at the back of the ramps. You'll find doing it a bit easier this time as you know what to expect when you bump against the stopping point. Just take your time and do it with confidence – no problem! Remember to check you have *no* area of wheel *overhang* off the sides of the ramps – peace of mind and *safety first*, friends!

Remove the ignition key, put the handbrake firmly on and place the car in first gear. If you place the car in park (for an automatic gearbox/transmission), it will only lock the two rear wheels. Once you're up on the ramps, follow the steps shown on Page 185 to 'Check the Car's Stability.' Carefully chock the front wheels and place your Detail Guardz on the outer edges of the front tyres. Remember to spread your tarp out under the car once again.

Refresh the Wash Water

Dump the dirty bucket water, give the bucket a quick rinse and refill it with clean hot water, mixed with two capfuls of Pure Shampoo plus two capfuls of APC. While you're at it, add some more hot water to the bucket of Waxoyl spray cans to ensure the Waxoyl will do the business when you need it.

Repeat Stages 1-4: Next, it's a case of again working through Stages 1-4 shown on Page 186-187 to detail under the rear half of the undercarriage. However, you don't need to repeat the Rust Repellent on the brake discs (rotors).

Stage 5: Dry the Undercarriage: Then, we need to give the whole undercarriage a thorough air blast with the Car Blower to expel most of the rinse water out of the nooks and crannies. Spend several minutes going over the whole underbody with the blower – when you think you've air blasted enough, give it another minute for luck!

To get the *best* from the Waxoyl, we need to go further with the drying activity. Grab your Portable Flatbed Fan Heater, put it on a *warm*, not hot setting, and slip it under the car. I used a small block of wood under the front feet to tilt the heat toward the underbody. The wee heater I used was the perfect size at just W24 x L24 x H11cm (W9.4 x L9.4 x H4.7in), bought on Amazon for £13.99 ($18.93). This thing was flat enough to move around under the underbody drying all areas with plenty of room above to spare.

Leave everything to dry for an hour or so while you have lunch or take a break; pop out every ten minutes (I set my phone alarm for this) to check drying progress and reposition the heater to a new drying spot. I found the heater perfectly safe to use in this way as it has an inbuilt thermostat that clicks off if it gets too hot - simples!

Safety first, friends – obviously, don't trail the power cable through any standing wash or rinse water!

Stage 6

Key to Successful Rattle Can Spraying - Shake It

The key to successful Waxoyl spraying is warming up the cans, which you've done. There must now be *frenzied* and prolonged rattle can shaking. After a bit of shaking, you'll hear the mixing ball start to rattle – keep going for a full two minutes shaking after you hear the rattle. Remember, you need to shake all five cans.

> ***Top Tip*** A Universal Paint Trigger makes Waxoyl spraying the undercarriage *oh so* much easier. See the Top Tip in Chapter 9 on Page 154.

When finished shaking the cans, securely clip your Universal Paint Trigger onto the collar at the top of the first rattle can.

If you're working in *cold* conditions, make sure the Paint Trigger is at room temperature *before* pushing it onto the rattle can collar. If the plastic around the Paint Trigger aperture is *too cold*, it will be brittle and may *crack* as you push it onto the rattle can – if that happens, the Paint Trigger will be useless.

Let's Get Waxoyl Spraying

When the underbody has had a good hour's drying, move the heater out of the way. Get down low on the mats on both sides and use your LED light to have a good look at the underbody to see what you're working to protect. It won't be immaculate under there, but your 4-stage detailing process will have *massively* improved the undercarriage and prepared it for the protection stage.

Now might be a good time to take a few snaps of the underbody, in case you want to show any potential future buyer your mint undercarriage, or just for your reference and enjoyment later!

PPE Reminder

I'm pretty sure I don't need to remind you to wear old clothes doing this; if Waxoyl lands on you, it will make a sticky mess! I also highly recommend wearing an old hoody or hat, plus Gloves, Eye Protection and a Face Mask. The Waxoyl will leave a thick, waxy textured finish on anything you spray it on or anything it lands on - which is what you want for protection purposes. I won't pretend the finish looks *glamorous*; it's not supposed to, but it does the job of protection, and remember, your shiny under trays will cover everything up when you've finished protecting under there.

Important note - avoid getting too much Waxoyl on any plastics or rubbers of the undercarriage, as the petroleum-based fluid in which the wax is suspended could dry these out. I also avoided the exhaust system as I didn't want the stench of burning wax when driving the car.

How Much Waxoyl to Use?

I allowed *one* can per *quarter* of the underbody working area, plus one can after final inspection. So, when you get down low, *visualize* the working zone quarter to be sprayed. I found that breaking up the area to be sprayed helped me get full coverage without too much waste by going over areas already done. Use the tape strip on the side of the car as a visual aid to help guide your spray zone separation.

A really helpful aspect of the Hammerite Waxoyl spray can is that it will pretty much work at *any* angle. This is a big advantage when working down low, reaching beneath the underbody into the tight spots.

First Waxoyl Quarter - Rear Half: Check that the spray can nozzle exit hole is central in the Paint Trigger aperture, so the edge of the aperture doesn't foul your nice smooth stream of Waxoyl spray and ruin your accuracy when you squeeze the trigger.

Be mindful that you want to get the Waxoyl anywhere where water is likely to collect. Give the can another quick shake. Working first on the rear raised half of the car, get down low on your mats again and reach under to start laying down the Waxoyl spray over your first working quarter of undercarriage – now you'll see why you needed the tarp! Try to focus the spray on the areas where you can see surface rust. Keep going on the first quarter until the can is empty.

Second Waxoyl Quarter - Rear Half: Move your Paint Trigger onto the second Waxoyl can and give the can another quick shake for luck. Move round to the other side of the car and follow the same method as before to get full coverage over your second quarter working area. Remember to focus the spray on areas where you can see surface rust. Keep going on the second quarter until the can is empty.

Time to Do the Front Half

Turn the Car: Now, turn the car around to raise the front of the car on the ramps. Remove your Detail Guardz and Wheel Chocks from the front wheels and put them aside. Then, carefully and slowly drive forward down the ramps and turn the car around. Follow the same process as before: *slowly* and *steadily* drive the car forward up the ramps until you feel the car bump *softly* against the top resting point at the back of the ramps. Once again, remove the ignition key, put the handbrake firmly on, and place the car in first gear. Carefully chock the two rear wheels that remain on the ground. Ensure your tarp positioning is still good and that the ramps are not on top of any part of the tarp. You shouldn't need your Detail Guardz again on this project.

Third Waxoyl Quarter - Front Half: Move your Paint Trigger onto the third Waxoyl can and give the can another quick shake. Get down low at the front and follow the same method as before to get full coverage over your third-quarter working area. Keep going on the third quarter until the can is empty. Remember to avoid getting too much Waxoyl on any plastics or rubbers of the undercarriage as best you can. It's impossible to miss them completely, so just stay aware.

Fourth Waxoyl Quarter - Front Half: Move your Paint Trigger onto the fourth Waxoyl can and give the can another quick shake. Get down low at the front on the other side of the car and follow the same method as before to get full coverage over your fourth quarter working area. Keep going on the fourth quarter until the can is empty.

Final Inspection

Get down low again, and go around the car, checking with your LED light that you haven't missed any patches of the underbody. If anywhere looks like it needs another wax hit, use your fifth and final can on any unprotected spots.

Touch Drying Time

Depending on the ambient conditions - temperature, humidity, etc., the Waxoyl should be touch dry 'tacky' in around 6-9 hours; at least it was on my Cayman in the June warmth.

Project Done!

You are *DONE* – congratulations, Driveway Warrior! I bet that undercarriage looks as *mint* as the freshly laundered under-crackers in the bedroom drawer! What a *massive* achievement – great job! The next task is to carefully reverse the car down off the ramps and park it away from the drive while you do a bit of clear up. The next thing I did was send a tired but happy 12-year-old niece indoors for a thorough decontamination in a hot bubble-filled bath prepared by her aunty. She seemed to have almost as much Waxoyl on herself as the undercarriage. Fortunately, she found this *hilarious* - children are such strange but wonderful creatures. Thank goodness I PPE'd her head to toe before we started!

I couldn't bring myself to try and save the £5.99 tarp, it was utterly destroyed, and any attempt to pressure wash it clean would just have transferred the waste Waxoyl off the tarp and onto my driveway. So, I stuffed the knackered tarp in a bin bag and binned it! On the bright side, the driveway was utterly immaculate – result!

Clear Up

Yep, time for one final clear up to close this book of epic detailing projects. I know you're tired, but let's push through and get it done - think of the fine icy fizz waiting in the fridge and pizza fest coming very soon! Start by returning all your Liquids and Spray bottles to storage, but do check all lids and spray heads are secure and keep an eye on any dilutables that need topping up.

Chuck the empty Waxoyl cans in the recycling. Grab your Car Blower, Caddy, Mats, Stool, Detail Guardz, Wheel Chocks and give them a quick wipe down to dry if necessary and store them away. Grab your Ramps - the chances are they'll be glazed with Waxoyl, so give them a rinsing Pressure Washer blast and leave them somewhere to dry before storing them away.

Discard all the wash bucket water and give your Wheel Bucket and Grit Guard a rinsing blast with the Pressure Washer. Hold back the Wheel Bucket for now, as you'll need it in a bit. Empty the spare bucket (rattle can warming water) in the water butt if you have one.

You should be left with your Pressure Washer, Extra-Long Handle Brush, Wheel Wash Bucket and Grit Guard. Dump the Brush in the Wheel Bucket, add a squirt of washing-up liquid, and quarter-fill it with hot tap water, swirling it around to mix. Spend a minute with your gloved hands in the soapy water, massaging the Brush clean. Dump the dirty soapy water and give the Brush a quick tap rinse. Then give it a good few wrist flicks to fling off the excess rinse water and put it somewhere to dry.

Wipe the Wheel Wash Bucket, Grit Guard and spare bucket dry and store them all away. Give the Pressure Washer and accessories, including the power cable, a wipe to dry and store them away. Finally, gather all remaining used Cloths, pop them in the washing machine with the day 1 batch, add the Microfibre Detergent and stick on a 40°C (104°F) 30-minute cycle with a gentle spin. When you take them out of the machine, give everything a good shake to re-set the fibres, and hang everything on a clothes dryer to dry naturally.

The next day when your Cloths and Brush are fully dry, store them away, keeping your Wheel Brushes in your Wheel Wash Bucket.

Don't Forget the Under Trays

Check the bag of under tray fixings is in the car, and carefully manoeuvre the five gleaming under trays into the back of your car. Steady now, even though they're clean, don't scuff them against the interior and leave a plastic scuff mark behind. Provided it's another dry day - you don't want to mess up your pristine undercarriage on wet roads, take a sedate trundle over to your mechanic to replace the trays. Go easy on the way there; remember your undercarriage isn't protected from flying road debris until the trays go back on. While the car was up on the lift, I took the chance to snap the undercarriage before the trays went back on.

In Retrospect

If I did this project again, I reckon I'd probably swap the Kneeling Mats for a mechanic's wheeled 'car creeper' to increase the speed and convenience of shuffling back and forth on your back while cleaning under the car. But I don't have a creeper, so the mats were an ok compromise. I've seen creepers on Amazon from around £33 ($44), but I can't recommend anything I haven't used myself.

Reflection

The brilliant thing about this project is you've just saved £525 ($710) on what the pro detailers quoted for the job; you've so *spectacularly* done yourself at a fraction of the cost! This was the average of three quotes from UK pro detailers.

So, Driveway Warrior, we *sadly* come to a close on this *epic* home detailing journey - it's been a hell of a ride, my friend! If you've come this far, I'm honoured and proud that you have done so. However, more importantly, YOU will have *transformed* your ride back to the *dazzling* beauty that graced the sales brochure – how *awesome* is that?

No doubt, there will have been some hiccups on the way, alongside the successes. But I'll wager you've emerged from your detailing adventures one *savvy,* confident and competent Driveway Warrior. With a *cracking* arsenal of kit and an even more *awe-inspiring,* impressive new set of knowledge and skills. You, Sir or Madam, are utterly *magnificent.*

I hope you'll kick back now, savour that delicious favourite pizza washed down with the fine icy fizz you've been saving. While you do that, please reflect on this *eye-watering* figure, which is the average cost that three UK professional detailers quoted for nine of the ten detailing projects in this book (excluding difficult to quote stone chips repair) - not a penny less than **£3,010 ($4,070)** – *ouch!*

Please look out for the next in the Driveway Warrior series, coming in the not-too-distant future. Meanwhile, why not head over to my blog at **driveway-detailing-warrior.com** and join me for all manner of home detailing fascination snippets? Why wait? See you there ...

The End

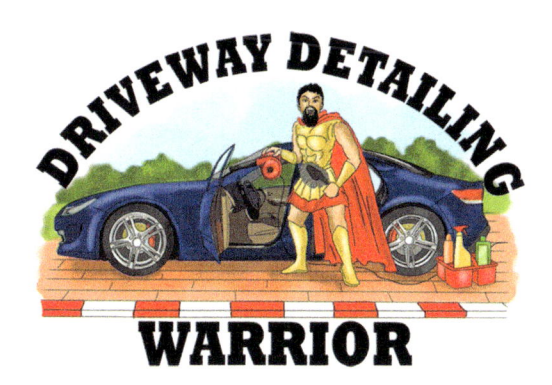

Dear Reader,

I sincerely hope you enjoyed reading my first book. I'm honoured and proud you have done so, and I thank you warmly, Sir or Madam, from the bottom of my heart. If you're able to post an honest review on the site you purchased the book from (no matter how short - a single line is perfectly fine), it would help me enormously and would be greatly appreciated.

Why not head over to my author website and blog now? The address is **driveway-detailing-warrior.com**, and you'll receive a cracking free bonus offer if you sign up to my free mailing list. I'll see you there!

Kind regards,

S. L. Lucas

Acknowledgements

For *me*.

For my wife Andrea, for somehow surviving this life with me by her side for over 11 years. For this, I nominate her for an award.

For my mum, for her saintly support in proofreading 64,000 words of material on a topic in which she has no interest *whatsoever*! I'm sorry for denting the oak bureau with my Lightsaber when I was ten - yes, it was me.

Simon, for his inspirational artwork depicting the *fearsome* Driveway Warrior in all his glory on which the cover is based.

Keith, the Master Mechanic at Golding Barn Garage, my friendly local independent Porsche Specialist, for patiently advising me on safe vehicle lifting and lowering.

I want to thank everyone who supported me in creating this book, except the smarmy Used Car Salesman in Chapter 8, who offered me five times less than I sold my car for. You, Sir, are an epic @$*!%#! I'll never forget that time we met, but I'll keep trying.

For Emma Watson, you seem like a lovely girl. Though you do linger in my thoughts, I'm happily married, so we ought just to be friends - *sorry*.

To my wonderful readers. The next book in the Driveway Warrior Adventures series is underway.

About the Author

S. L. Lucas is the owner and creator of the popular *'Driveway Detailing Warrior'* home detailing blog and is a blogging mentor in the home detailing community. A published writer for the local press as well as a Porsche monthly magazine, he has bought, sold, detailed, cherished and spiritedly driven three different Porsche models since 2002. He lives in Brighton and continues to detail and enjoy his current cherished Porsche Cayman 987.2.

Back Pages Marketing Bits

Interested in buying ten or more copies of this book?
For discounted bulk purchases, please email us at

driveway-warrior@outlook.com

To book S Lucas for interviews or speaking engagements, please email us at

driveway-warrior@outlook.com

Connect with me on Facebook: Sean Lucas (Driveway Detailing Warrior)

Printed in Great Britain
by Amazon